Artificial Intelligence Artificial Communication

Artificial Intelligence Artificial Communication

Tamer Bayrak

PETER LANG
Berlin · Bruxelles · Chennai · Lausanne · New York · Oxford

Library of Congress Cataloging-in-Publication Data

Names: Bayrak, Tamer, 1967- author.
Title: Artificial intelligence artificial communication / Tamer Bayrak.
Description: New York : Peter Lang, [2025]
Identifiers: LCCN 2025004116 (print) | LCCN 2025004117 (ebook) | ISBN 9783631930885 (paperback) | ISBN 9783631936078 (ebook) | ISBN 9783631936108 (epub)
Subjects: LCSH: Telematics. | Digital media--Social aspects. | Chatbots--Social aspects. | Social influence. | Social media. | Artificial intelligence. | Mass media--Influence. | Authorship--Data processing.
Classification: LCC TK5105.6 .B39 2025 (print) | LCC TK5105.6 (ebook) | DDC 006.3/5--dc23/eng/20250304
LC record available at https://lccn.loc.gov/2025004116
LC ebook record available at https://lccn.loc.gov/2025004117

Bibliographic Information published by the Deutsche Nationalbibliothek
The Deutsche Nationalbibliothek lists this publication in the Deutsche Nationalbibliografie; detailed bibliographic data is available in the internet at http://dnb.d-nb.de.

ISBN 978-3-631-93088-5 (Print)
E-ISBN 978-3-631-93607-8 (ePDF)
E-ISBN 978-3-631-93610-8 (ePUB)
DOI 10.3726/b22796

© 2025 Peter Lang Group AG, Lausanne (Switzerland)
Published by Peter Lang GmbH, Berlin (Germany)

info@peterlang.com

All rights reserved.

All parts of this publication are protected by copyright.
Any utilization outside the strict limits of the copyright law, without the permission of the publisher, is forbidden and liable to prosecution.
This applies in particular to reproductions, translations, microfilming, and storage and processing in electronic retrieval systems.

This publication has been peer reviewed.

www.peterlang.com

To My Beloved Wife Eda …

Contents

List of Figures viii

Preface ix

    Introduction 1

1 Foundations of Artificial Intelligence 23

2 Artificial Communication Tools and Technologies 43

3 Ethical and Societal Impacts 57

4 The Use of Artificial Intelligence in Communication Sciences 77

Conclusion and General Evaluation 135

Bibliography 139

List of Figures

4.1	ChatGPT's workflow	109
4.2	Example of general prompt structure	123
4.3	Prompt production model in communication studies	127

Preface

In today's rapidly advancing technological landscape, the concepts of artificial intelligence and artificial communication have captured the attention not only of scientists but also of entire societies. This study examines the foundations of artificial intelligence and artificial communication, along with recent developments and their areas of application, focusing on how they impact and transform human life. Emphasis is placed on the innovations brought about by key technologies such as machine learning, deep learning, and natural language processing, while also addressing the social, economic, and ethical dimensions of these technologies.

Initially perceived as a concept confined to science fiction films, artificial intelligence has now become an indispensable part of our daily lives. This study presents a historical overview of the development of artificial intelligence and artificial communication tools, along with their theoretical foundations and projections of what the future holds. While artificial communication transforms human-machine interaction, it also reshapes societal structures, aiming to explore the intersection of social sciences and technology in this context.

I hope this study will shed light on the future of artificial intelligence and artificial communication, serving as a guide for researchers and students working in this field. I would also like to express my gratitude to my colleagues, my family, and my beloved wife Eda, who have constantly supported me throughout this journey.

Sincerely,
Tamer Bayrak
2025

Introduction

What is Artificial Intelligence?

Artificial intelligence (AI) refers to a type of intelligence that involves the creation of systems capable of reasoning, learning, and acting autonomously. Research conducted on AI has introduced effective techniques designed to solve various issues concerning humans and society (Russell & Norvig, 2010, p. 28).

AI is a rapidly evolving field that will undoubtedly have a profound impact on both individuals and society in the coming years. It will be utilized in various sectors, from driverless cars to pilotless airplanes, and from the energy industry to the construction sector (Brynjolfsson & McAfee, 2014, p. 10). Even today, products associated with this type of intelligence have begun to emerge. As AI technology advances, its influence on all tools integrated into modern human life will become increasingly evident (Tegmark, 2017, p. 30).

It cannot be said that there is significant resistance to the development of AI today. The primary reason for this is the societal acceptance of new media technologies. People are no longer surprised by such technologies. Driverless vehicles, unmanned aerial vehicles, macro- and micro-operating tools in the space domain, autonomous robots performing daily tasks, and many others have become commonplace for individuals. Therefore, the societal resistance previously experienced during the development of new technologies is not anticipated. For instance, the nearly 20-year journey of the internet becoming ordinary in everyday use might be the last and longest-developing technology within new media technologies (Castells, 2010, p. 386).

AI is an intricate and fast-evolving discipline that cannot be easily summarized. Fundamentally, it involves the capability of machines to carry out tasks autonomously that would typically need human input (Russell & Norvig, 2010, pp. 28–29). In this context, the abilities to reason, rapidly learn, solve problems, increase efficiency, and provide recommendations are key expectations (Tegmark, 2017, p. 150).

To meet these expectations, AI requires algorithms. An AI algorithm consists of a set of instructions that tells a computer system how to behave (Murphy, 2012). One of the most significant examples of this is the Naive Bayes classifier. This classifier employs a probabilistic approach and is used for pattern recognition problems in scenarios with restrictive premises (Domingos & Pazzani, 1997, pp. 105–106). Although its use comes with inherent limitations, the Naive Bayes classifier can yield effective results by incorporating methods such as artificial neural networks, a type of information processing technology similar to the human brain's processing technique, thanks to its flexibility in statistical analysis. However, it can also clearly reveal the disadvantages of artificial neural networks. For instance, what happens within the system may be unknown, stability analysis and testing may not be feasible outside specific networks, and difficulties may arise in using it across different operating systems.

The Naive Bayes classifier, a widely used machine learning algorithm for classification problems, is derived from Bayes' theorem, first introduced by English statistician Thomas Bayes (Koller & Friedman, 2009, p. 13). In this context, it assumes that the features in the input data are conditionally independent of one another (Mitchell, 1997, p. 180). Bayes' theorem is considered within the context of probability theory. The theorem demonstrates the relationship between conditional and marginal probabilities in the probability distribution of random variables. In this form, Bayes' theorem specifies relationships that are acceptable to all statisticians. According to the probability stated in the theorem, the probability value of event A given the conditions of event B differs from the probability value of event B given the conditions of event A. Additionally, there is a definite relationship between such probabilities.

Bayesian classifiers can also be considered as Bayesian networks, where each feature is conditionally independent of the others, and the concept to be learned is conditionally dependent on all these attributes (Pearl, 1988, p. 116). A Bayesian network is also referred to as a non-regressive graphical model (a branch of mathematics that studies graphs). In this framework, it is defined as a probabilistic graphical model (a type of statistical model), a form of statistical model, which illustrates a group of random variables along with the conditional dependencies that exist among them. Bayesian networks are an efficient modeling type for describing an event that occurs in daily life and predicting how any one of the known possible causes of that event might contribute to the probabilities of the factors involved (Koller & Friedman, 2009, p. 45).

The probabilistic relationships between cause and effect for any event can be modeled using a Bayesian network. For example, using this methodology, the probability of a person having a specific disease can be calculated based on the symptoms they exhibit (Jensen, 1996). All events involving cause-and-effect relationships, like this and others, can be explained both qualitatively and quantitatively. Quantities, variables, parameters, and hypotheses to be tested can be considered as Bayesian probability variables (Koller & Friedman, 2009, pp. 12–14). In this context, units that do not intersect in any way are assumed to be conditionally independent of each other. The existing units are associated with a probability function that takes the values of the upper layers as input and generates probability values for the possible values that the variable represented by the relevant unit can assume.

There are effective inference and learning methods that use Bayes' theorem. By classifying a collection of variables, Bayes' theorem serves as a significant theorem in mathematical statistics due to this characteristic (Pearl, 1988). The purpose of this theorem is to reach a conclusion by processing comprehensive facts, as well as qualitative and quantitative observations, to model a random phenomenon (Gelman et al., 2013). Compared to classical statistical methods, this approach presents ambiguous data in the form of observation, interpretation, and classification. In Bayesian statistics, probability is inductive probability. The main goal here is to conclude the experiment in a deterministic manner with the highest probability (Bernardo & Smith, 2000, pp. 16–33). Bayesian methods examine probability theory from a logical perspective and proceed by evaluating the existing probabilities. Particularly, the values to be counted take values between 0 and 1 in a format that facilitates quantitative counting in a computer environment, and in this context, the hypothesis is tested.

Supervised learning, clustering, dimensionality reduction, structured prediction, anomaly detection, neural networks, and reinforcement learning, as part of machine learning and data mining, form a complex and multi-layered framework. An important example of AI within this context, Naive Bayes stands out as a simplified version of Bayes' theorem, accounting for the assumption of independence (Mitchell, 1997, p. 154). The classification of this approach is part of the supervised machine learning class (Domingos & Pazzani, 1997, pp. 103–130). More specifically, the classes that need to be recognized and the classes to which the sample sets are associated are different. An example of this is the procedure for recognizing and blocking spam emails. Spam and non-spam emails are considered the two email classes to be analyzed. An algorithm that determines whether an email is spam is an example of supervised machine learning. In this context, during the classification process, a typical model exists. The procedure at hand involves dividing the model into predefined classes. All patterns are characterized by a set of values.

When this theorem is explained with an example, its application becomes somewhat clearer. As an example of the application of the Naive Bayes theorem, spam email filtering can be given. This method can be used to determine whether an email is spam. The necessary steps should be classified into five stages: preparing the dataset, identifying features, calculating probabilities, applying Bayes' theorem, and making a decision.

Preparing the Dataset: First, a labeled dataset is collected. This dataset contains emails classified as "spam" and "not spam."

Identifying Features: Words in the emails are used as features. For example, words like "free," "win," and "now" are identified.

Calculating Probabilities: The likelihood of each word occurring in spam emails is determined, and similarly, the likelihood of each word appearing in non-spam emails is also assessed.

Applying Bayes' Theorem: Bayes' theorem is used to calculate the probability that a new email is spam. The probabilities of the words in the email being spam are multiplied to find the total probability.

Decision-Making: Based on the calculated probability, the email is classified as either "spam" or "not spam."

For example, if an email contains the words "free" and "win," and the probabilities of these words being associated with spam are high, the email is classified as spam. In this context, if a calculation is to be made:

Dataset: 1,000 emails (500 spam, 500 non-spam).
Features: "free," "win," "now."

Probabilities:

The word "free" appears in 300 spam emails and 50 non-spam emails.
The word "win" appears in 250 spam emails and 30 non-spam emails.
The word "now" appears in 200 spam emails and 20 non-spam emails.

Application of Bayes' Theorem:

The probability of spam: P(Spam|free) = (300/500) * (500/1000) / (350/1000)

Bayes' Theorem:

P(Spam|free) = P(free|Spam) * P(Spam) / P(free)

Step 1: Calculation of Probabilities
- P(Spam) = 500 / 1,000 = 0.5
- P(free|Spam) = 300 / 500 = 0.6
- P(free|Not Spam) = 50 / 500 = 0.1
- P(free) = (300 + 50) / 1,000 = 0.35

Step 2: Application of Bayes' Theorem
P(Spam|free) = 0.6 × 0.5 / 0.35 = 0.3 / 0.35 ≈ 0.857

Conclusion: If the email contains the word "free," the probability of it being spam is approximately 85.7 percent.

This simple example illustrates the process by which the Naive Bayes classifier determines whether an email is spam. If the email contains more words, similar probability calculations can be performed for each word, and the overall probability can be determined by multiplying these values. In this way, the Naive Bayes classifier is able to calculate the likelihood of an email being spam with high accuracy, based on the presence of specific words.

History and Development

AI can be defined as the creation of machines that exhibit capabilities similar to human intelligence. The history and development of AI are based on several key milestones, with numerous achievements and challenges experienced throughout the progression of this field. The history of AI encompasses various distinct periods from its inception to the present day.

Early Period and Foundations

The origins of AI date back to the mid-twentieth century, parallel to the development of computer science. In the 1950s, Alan Turing's work addressing the question "Can machines think?" laid the foundation for AI research. Turing proposed the idea that machines could mimic human intelligence and developed the Turing Test, which is designed to evaluate a machine's ability to demonstrate thinking abilities (Russell & Norvig, 2010, pp. 16–18). In his paper "Computing Machinery and Intelligence," Turing proposed this test to determine whether a machine could exhibit human-like intelligence. The test is based on the idea that if a human interrogator, engaged in a written conversation with both a computer and a human participant, is unable to distinguish which one is the computer, then the computer is considered to be "thinking" (Turing, 1950, p. 433).

Modern evaluations and applications of the Turing Test have evolved alongside the development of AI. For example, modern AI systems like AlphaZero and AlphaStar, developed by DeepMind, have achieved significant success by playing complex strategy games at a superhuman level. The success of these systems has raised new questions about how AI can perform in real-world applications, going beyond the original scope of the Turing Test (Damassino, & Novelli, 2020, pp. 463–468).

There are also updated versions of the Turing Test. For example, IBM's Project Debater is used to evaluate whether AI can demonstrate human-like abilities in a debate format. Such tests aim to measure not only the computational capacity of AI but also its ability to communicate and develop arguments, skills typically associated with human intelligence (Modernizing The Turing Test For 21st Century AI, 2021).

The Emergence of the Term Artificial Intelligence

The term "artificial intelligence" was first used by John McCarthy in 1956 during the Dartmouth Conference. This conference played a significant role in the recognition of AI as an academic discipline. During the conference, McCarthy and other participants discussed the methodologies necessary for developing machines capable of learning and problem-solving. (Russell & Norvig, 2010, p. 17). Key figures such as John McCarthy, Marvin Minsky, Nathaniel Rochester, and Claude Shannon participated in this conference, laying the foundations of AI. Following the Dartmouth Conference, AI research rapidly progressed. In addition to McCarthy, researchers like Allen Newell and Herbert A. Simon made significant contributions during this period. Newell and Simon developed a program called "Logic Theorist," one of the first AI programs. This program had the ability to prove logical theorems, demonstrating the potential of this field (McCorduck, 2004).

Early Successes and Challenges

From 1957 to 1974, the field of AI experienced significant growth. As computers became capable of storing more information, became faster, cheaper, and more accessible, and as machine learning algorithms advanced, AI research gained momentum. Early successes such as Newell and Simon's General Problem Solver and Joseph Weizenbaum's ELIZA program demonstrated the potential of AI (The History of Artificial Intelligence, 2023). However, the greatest challenge during this period was the lack of processing power and the limited computational capacities of computers.

The First Winter and Revival

In the 1970s, AI research entered a significant period of stagnation. During this time, due to the lack of sufficient computing power and the overly high expectations, funding decreased, and research slowed down. This period is known as the "first AI winter." However, in the 1980s, AI experienced a revival with the popularization of deep learning techniques by John Hopfield and David Rumelhart. During this time, Edward Feigenbaum's expert systems also gained widespread use (Russell & Norvig, 2010, pp. 22–25). During the revival period, the focus shifted toward new approaches such as machine learning and artificial neural networks.

In the 1980s, AI research saw a revival, driven by two key factors: the development of advanced algorithmic methods and a boost in financial support. Deep learning approaches, introduced by John Hopfield and David Rumelhart, allowed machines to improve through experience. By the late 1980s, expert systems had become widely used across various industries.

However, the success of these systems was limited, and the anticipated revolutionary outcomes were not achieved. Despite the failure of projects like the Fifth Generation Computer Project (FGCP), the engineers and scientists trained during this period made significant contributions to the advancement of AI in subsequent years.

Social and Economic Impacts

The development of AI has not been limited to technical advancements; it has also had significant social and economic impacts. Technologies such as big data analytics, automation, and robotic systems have transformed production processes and led to major shifts in the labor market. In this context, the economic effects of AI have been profound, contributing to increased productivity, the emergence of new business models, and economic growth. Big data analytics and automation have enhanced efficiency in production processes and reduced costs. These technologies have enabled companies to manage their operations more effectively and gain a competitive advantage. Furthermore, the innovative solutions offered by AI have allowed firms to better understand customer needs and respond more quickly in the marketplace (Haenlein & Kaplan, 2019, pp. 6–8).

The economic impacts of AI are observed across a wide spectrum, from productivity increases to shifts in the labor market. In particular, AI technologies have been identified as significant contributors to efficiency gains. For instance, Julius Tan Gonzales' study titled "Implications of AI Innovation on Economic Growth: A Panel Data Study" aimed to measure the impact of AI on the economy, especially its long-term effects on economic growth. The study hypothesized a positive relationship between AI and economic growth. At the conclusion of the research, a positive relationship between AI and economic growth was found, with additional findings indicating that the impact of AI on growth is more robust among developed economies (Gonzalez, 2023).

The effects of AI on the labor market are complex and multidimensional. In some sectors, automation may reduce the demand for labor, potentially increasing unemployment rates, while in other sectors, it contributes to the creation of new job opportunities. In this context, the study by Brynjolfsson and McAfee, which examines the transformations in the labor market brought about by AI and automation technologies and their economic impacts, is particularly significant (Brynjolfsson & McAfee, 2014). Accordingly, the widespread adoption of automation, particularly in low-skilled jobs, increases the risk of job displacement for workers in these roles, while AI-supported new job opportunities are being created in high-skilled occupations. Moreover, the societal impacts of AI are also extensive. While AI technologies may contribute to increasing social inequalities, they can also enhance quality of life by improving education, healthcare, and other social services. As demonstrated in Autor's study, the transformations in the labor market due to automation and AI technologies, and their economic impacts, are particularly noteworthy (Autor, 2015). Particularly, the widespread adoption of automation in low-skilled jobs increases the risk of job displacement for workers in these roles, while AI-supported new job opportunities are being created in high-skilled occupations.

The societal impacts of AI are also extensive. While AI technologies may contribute to increasing social inequalities, they can also enhance quality of life by improving education, healthcare, and other social services. In the study by Korinek and Stiglitz, the societal impacts of AI technologies and their potential effects on social inequalities are discussed (Korinek & Stiglitz, 2017). Evaluations of the ethical and societal impacts of AI technologies help us understand how these technologies affect the social structure. In this respect, as emphasized at the core of the study, such research is of great importance for discussing how technologies that create superhuman intelligence levels may influence inequality.

Its Importance Today

AI is leading to significant advancements in many fields today and plays a key role in the transformation of these areas. The opportunities presented by AI, along with the challenges it faces, are reshaping societal and economic structures.

In this context, AI makes significant contributions to economic growth and productivity. It plays a major role, particularly in increasing efficiency in business processes, reducing costs, and accelerating innovation. In this regard, Brynjolfsson and McAfee (2014) highlight the innovations and economic advantages brought by AI in the business world. AI's role in the labor market, especially through the automation of low-skilled jobs, increases unemployment rates, while creating new opportunities in high-skilled positions. In the healthcare sector, AI applications enable groundbreaking advancements in areas such as early diagnosis of diseases, treatment planning, and patient monitoring. The use of AI in healthcare enhances patients' quality of life and improves the efficiency of healthcare services. Studies in this field clearly demonstrate that AI increases efficiency and accuracy in healthcare services. For instance, in the study by Shen et al., the effectiveness of AI in helping doctors obtain both qualitative and quantitative data in a repeatable manner for disease detection was discussed (Shen et al., 2021). Similarly, the comprehensive literature review conducted by Kitsios and colleagues emphasizes the role of AI in early diagnosis and diagnostic processes, examining the various advantages it provides in healthcare services. Additionally, the social and ethical impacts of AI on individuals, healthcare professionals, and the healthcare industry are evaluated, highlighting its benefits (Kitsios et al., 2023). Furthermore, the study by Chaddad and colleagues, which focuses on explainable AI (XAI) techniques used in healthcare services and medical imaging applications, is noteworthy (Chaddad et al., 2023). The study summarizes how deep learning models can be made more interpretable and which algorithms can be used to enhance the reliability of these models in medical decision-making processes. It also provides suggestions on how XAI concepts could guide future research in medical text and image analysis. As can be understood from this and similar studies, the search for AI-supported solutions in the healthcare field is accelerating. However, it is evident that the focus of these studies is on enhancing the supportive qualities of AI in assisting doctors with manual disease detection. In other words, the aim is not to replace doctors, but rather to support and shorten the diagnostic process.

AI also offers significant innovations in the field of education. AI applications are used in areas such as monitoring student performance, creating personalized education programs, and increasing access to educational materials. AI-supported educational technologies enhance equity and access in education by offering customized learning experiences tailored to students' individual needs. AI algorithms, which enable personalized learning experiences by providing educational materials and programs adapted to students' individual needs and learning pace, can analyze students' strengths and weaknesses and recommend the most effective materials and methods to facilitate learning.

AI also provides substantial support to teachers, allowing them to closely monitor student performance and intervene when necessary. These technologies enable teachers to track student progress and prepare more effective lesson plans tailored to individual needs. Furthermore, the data provided by AI helps teachers identify areas where students are struggling or excelling. In this context, AI-supported tools play a crucial role in improving the quality and accessibility of educational materials, offering students access to a wide range of high-quality educational resources. This is particularly facilitated by advanced AI systems' translation and customization capabilities, overcoming language and cultural barriers to provide a global learning experience. For example, language learning applications help students develop their language skills by making educational materials available in various languages.

AI is also used in learning analytics and feedback processes. Learning analytics utilizes data mining and machine learning techniques to monitor and analyze students' learning processes. This allows the identification of areas where students are struggling or excelling. Teachers can use this data to provide more effective feedback to students and optimize their learning processes (Holmes et al., 2019).

AI-supported educational technologies make learning processes more efficient, effective, and accessible. In the future, AI is expected to further advance in the field of education, offering systems that focus more on students' individual needs and enhance equal opportunities in education. Research shows that the applications of AI in education are continuously expanding and improving (Salas-Pilco et al., 2022, pp. 1–3).

Social and Ethical Impacts

The social and ethical impacts of AI are also quite extensive. Issues such as data privacy, algorithmic bias, and ethical use have become significant topics of discussion with the widespread adoption of AI. Studies in these areas help us understand the ethical and social dimensions of AI technologies and facilitate the implementation of necessary regulations to ensure these technologies are used in a fair and transparent manner (Floridi & Cowls, 2019, pp. 8–9).

The current significance of AI has been shaped by technological advancements and scientific discoveries, and it will continue to drive transformation in many areas in the future. The advantages offered by AI, as well as the challenges it faces, contribute to the reshaping of societal and economic structures. In this context, the AI principles established by the United Nations Educational, Scientific and Cultural Organization (UNESCO) highlight the importance of AI and demonstrate that it is one of the most crucial technological developments shaping the future of humanity. What is certain is that, in its various forms, AI will impact and transform human life, the environment, and nature. By adhering to the fundamental ethical principles set and recommended by UNESCO, minimizing the risks of AI while maximizing its potential benefits is a responsibility shared not only by those developing the technology but also by its users. The sustainable and equitable use of AI technologies can only be achieved through the collective awareness created by both producers and consumers (UNESCO, 2021).

The fundamental principles put forward by UNESCO are as follows:

Proportionality and Do No Harm: AI must provide equal access and ensure this capability for all social groups. It should promote social equality by allowing access for marginalized groups and facilitate access to information. In this context, AI systems should not be used beyond what is necessary to achieve legal purposes. A risk assessment must be conducted to prevent dangers arising from such use.

Safety and Security: AI must respect human rights and work to positively impact the quality of human life. In this regard, technology should be developed with ethical principles in mind, and these principles should be prioritized in its use. AI actors must avoid and address accidental harm, and they must mitigate security risks and vulnerabilities.

Right to Privacy and Data Protection: The procedures for operating AI systems and algorithms should be clearly explained. Those who wish to use this technology must be able to easily understand the principles of AI and the data being used. Privacy should be guaranteed and supported throughout the entire AI lifecycle as much as possible. Adequate data protection frameworks should be established and maintained.

Multi-Stakeholder and Adaptive Governance and Collaboration: Individuals developing and using AI technology must take responsibility for the impacts of AI. They should absolutely avoid negative outcomes, take necessary precautions, and correct any identified errors. Additionally, international law and national autonomy should be respected in data governance. The involvement of various stakeholders is essential for inclusive approaches to AI regulation.

Responsibility and Accountability: While AI must respect users' privacy and data protection rights, it should also store data effectively and securely. Effective enforcement measures against data crimes must be implemented. In this regard, AI systems should be traceable and auditable. Oversight, impact assessment, auditing, and due diligence should be integrated into the mechanism to prevent conflicts with human rights principles and risks to environmental health.

Transparency and Explainability: AI systems must have a reliable architecture. They should be protected from cybersecurity risks, and preventive measures against malicious use must be taken. While doing so, transparency and explainability should not be overlooked. The ethical behavior of AI systems depends on their transparency and explainability. Since there can be tensions between these qualities and other principles like privacy, safety, and security, the level of transparency and explainability must be context-dependent.

Human Oversight and Determination: The scale, efficient use, and methods of resources must be shaped under human supervision. In this respect, the development and use of AI technology should be carried out under human oversight and determination, with consideration for the environment and natural resources. States using this technology must ensure that AI systems do not undermine human responsibility and accountability.

Sustainability: AI should contribute to long-term goals. With its important role in addressing environmental and climate change issues, AI can help create a livable world for future generations. AI technologies should be evaluated based on their impact on sustainability, which is a continuously evolving set of goals listed in UNESCO's sustainable objectives.

Awareness and Literacy: As AI technologies expand among users, it is increasingly important for people to possess digital knowledge and skills. Educational resources should be accessible to all segments of society. To improve public understanding of AI and data, efforts should focus on providing accessible and open education, fostering community involvement, enhancing digital skills, promoting AI ethics, and utilizing media and information platforms.

Fairness and Non-Discrimination: The development of AI should support communication between knowledge providers and stakeholders. For this, governments, funders, and scientific organizations must work together. AI actors should adopt an inclusive approach that ensures the benefits of AI are shared by everyone, promoting social justice, equality, and non-discrimination.

UNESCO and Artificial Intelligence Studies

UNESCO identifies AI as one of the central issues in current and near-future technological development. According to UNESCO, the impact of AI technology will depend on humanity's ability to control and manage it. In this context, UNESCO believes that as AI technologies continue to evolve, an ethical dimension must accompany this progress. Transferring bias and unfair preferences into technology during AI development will lead to the continuation and deepening of existing problems. Additionally, to protect personal data associated with AI use, ensure equal access to technology, and prevent the growth of inequality through AI, UNESCO has initiated the Standardization of AI Ethics. A recommendation on AI has been formulated, and principles have been prepared and shared.

The rise of AI has led to numerous opportunities in the global community, including diagnosing health problems, fostering human connections via social media, and enhancing workforce productivity through automated processes. However, these changes are accompanied by profound ethical concerns. The dangers associated with AI are beginning to have a greater impact on marginalized populations, adding to existing inequalities.

UNESCO has adopted an approach that emphasizes the unprecedented need for ethical principles in no other field as much as in this one. These general-purpose technologies are transforming the way we work, communicate, and live. The world is poised for a dramatic change, unseen since the introduction of the printing press over a century ago. While AI technology offers countless benefits, without ethical guidelines, it can reproduce real-world biases and discrimination, lead to division, and pose threats to fundamental human rights and freedoms.

In November 2021, UNESCO established the first official standards regarding AI ethics, called the "Recommendation on the Ethics of Artificial Intelligence." This framework was adopted by all 193 member states. The recommendation is grounded in the development of fundamental principles, such as the protection of human rights and dignity, transparency, and justice, and it consistently emphasizes the necessity of human oversight in AI systems.

Key policy areas highlight clear domains where member states can take steps to ensure the responsible development of AI. While values and principles are critical for laying the foundation of any AI ethics framework, recent advancements in AI ethics emphasize the need to move beyond high-level principles toward pragmatic strategies.

UNESCO has also developed incentive plans to encourage member states to take action in implementing these recommended principles. It is acknowledged that a long road lies ahead in terms of securing the necessary resources. In this regard, UNESCO has outlined two practical approaches. The first is referred to as a readiness assessment methodology. This methodology aims to facilitate the evaluation of member states' preparedness for effectively implementing the recommendations. It will help them recognize their levels of readiness and provide a foundation for UNESCO to focus its capacity-building efforts on specific areas. The second approach is the ethical impact assessment. This assessment is a systematic procedure designed to facilitate the alignment of AI projects with the communities affected by them. The process involves identifying and evaluating the impacts of AI, with the aim of facilitating reflection on the potential consequences of a program and determining necessary actions to prevent harm.

When examining the principles set forth by UNESCO, it becomes evident that AI technology is expected to develop rapidly. However, it is also emphasized that serious measures must be taken to prevent uncontrolled development or the loss of control, a general fear associated with this technology.

In this context, UNESCO's latest work on AI is also noteworthy. The Global Forum on AI Ethics held in Kranj, Slovenia, on February 5–6, 2024, the Massive Open Online Course (MOOC) on AI Ethics developed in collaboration with LG AI Research, the AI Readiness Assessment aimed at aligning Mexico's national AI strategy with ethical standards, and the establishment of the first UNESCO-affiliated AI center in Africa all demonstrate the ongoing active work in the field of AI. These initiatives are significant as they reflect UNESCO's continued commitment to promoting ethical AI practices and supporting global cooperation in AI governance.

The widespread adoption of AI technologies has brought with it the capacity to collect and process vast amounts of data. This raises significant concerns regarding data privacy and security. Accordingly, academic studies in this area have gained great importance. For instance, the study conducted by Zawacki-Richter et al. highlights the impact of AI applications used in the field of education on the privacy of student data. The study emphasizes the need for transparency and security measures in how student data is collected, processed, and stored (Zawacki-Richter et al., 2019, p. 2).

Moreover, AI systems can learn biases from data sets and reflect these biases in a skewed manner within decision-making processes. This can lead to an increase in social inequalities and discrimination. How algorithms reinforce societal biases and what measures should be taken to mitigate these biases have become important areas of research. Studies in this regard emphasize the need for AI systems to be fair and impartial (Binns, 2018).

In studies examining the social and ethical dimensions of AI, a framework is generally proposed based on five fundamental principles: beneficence, non-maleficence, autonomy, justice, and transparency. These principles provide guidance for the ethical use of AI technologies. Furthermore, it is emphasized that AI policies and regulations need to be developed at the international level (Floridi & Cowls, 2019, pp. 4–5).

However, if this is not done, there is a risk that existing inequalities could deepen further. For example, in regions without access to high-quality educational technologies, the digital divide may widen. To prevent this, careful consideration must be given to what is taught and how it is delivered (Holmes et al., 2019). From this perspective, understanding and managing the social and ethical impacts of AI is crucial to ensuring the more equitable and responsible use of these technologies. Therefore, the research conducted, and the proposed ethical frameworks provide essential guidance in maximizing the societal benefits of AI while minimizing its negative effects.

What is Artificial Communication?
Definition and Conceptual Framework

Artificial communication has permeated many areas of our lives with the advancement of modern technology and has become a part of our daily interactions. This type of communication refers to forms of communication that are generated, directed, or processed by machines or computer systems. Artificial communication can involve interactions between humans and humans, humans and machines, or machines and machines. This has been made possible particularly through the development of technologies such as AI, natural language processing (NLP), and machine learning. As Russell and Norvig have also noted, these technologies form the fundamental components of artificial communication (2010, p. 26).

AI and NLP are among the most important elements of artificial communication. These technologies enable computers to understand, interpret, and generate human language. For instance, digital assistants like Siri and Google Assistant can understand users' natural language commands and respond accordingly. These systems assist users with daily tasks such as answering questions, setting reminders, or playing music. This demonstrates how deeply human-machine interaction is integrated into our daily lives. NLP technologies are also used to overcome language barriers. Translation applications make communication in different languages possible, enhancing global interaction.

Another important application area of artificial communication is its supportive role for individuals suffering from neurological disorders. People affected by neurological conditions, such as the late astrophysicist Stephen Hawking, who lost the ability to use his voice due to Amyotrophic Lateral Sclerosis (ALS), are able to communicate through the use of AI and NLP technologies (Jurafsky & Martin, 2023, p. 338). These systems can detect signals such as brainwaves or muscle movements from users and convert them into meaningful language outputs. In this way, individuals suffering from such conditions are granted the freedom to communicate.

Machine-to-machine communication is also a significant dimension of artificial communication. This type of communication is particularly important in fields such as the Internet of Things (IoT) and Industry 4.0. The exchange of data between different machines or devices enables the automation of processes and increases efficiency. For example, smart home systems make it possible to manage interconnected devices through a central control unit. This allows users to save energy, enhance security, and improve their quality of life.

In addition, artificial communication plays a significant role in customer service and marketing. Chatbots and virtual assistants are widely used to respond to customer inquiries, resolve issues, and provide product or service recommendations. This capability helps businesses enhance customer satisfaction and reduce operational costs. Moreover, these technologies contribute to the development of personalized marketing strategies by analyzing customer data.

Differences Between Natural Communication and Artificial Communication

Natural communication encompasses conversations, written exchanges, and other forms of interaction that occur between people. In natural communication, linguistic and non-linguistic cues, context, and cultural norms are of great importance. In contrast, artificial communication involves messages generated and processed by machines, typically relying on algorithmic processes. Natural communication is enriched by emotional and social contexts, fostering deep understanding and empathy between individuals. On the other hand, artificial communication is more structured, logical, and data driven. The effectiveness of artificial communication depends on the quality of the algorithms and data sets used (Mitchell, 1997, p. 81).

Another difference between natural and artificial communication is the response speed. Natural communication is typically instantaneous and dynamic, whereas artificial communication can occur at varying speeds depending on the data processing capacity of the systems. Additionally, misunderstandings and inconsistencies in natural communication can stem from context and social norms, while errors in artificial communication are usually caused by algorithmic flaws or data deficiencies (Russell & Norvig, 2010, p. 888).

In addition to the difference in response speed, the sources of errors also differ. In natural communication, misunderstandings and inconsistencies typically arise from context and social norms. For instance, a person's statement might be misunderstood by another due to nuances in the spoken language or cultural differences. On the other hand, errors in artificial communication are usually caused by algorithmic flaws or data deficiencies. AI systems require accurate and comprehensive data sets, and any gaps or errors in these data can lead to incorrect outcomes. Furthermore, the complexity and accuracy of algorithms directly influence the performance of the systems. A flaw or deficiency in an algorithm can negatively affect the accuracy and reliability of the system's responses.

In conclusion, these differences between natural and artificial communication help us understand the advantages and disadvantages of both forms. While natural communication ensures instantaneous and contextual understanding, artificial communication offers structured and logical responses depending on the quality of data and algorithms. Therefore, when developing artificial communication systems, great importance must be placed on both the quality of data sets and the accuracy of algorithms.

Usage Areas and Examples

Artificial communication is used in many areas and has become a part of everyday life. For instance, digital assistants like Siri, Google Assistant, and Alexa, frequently used in new media environments, understand users' natural language commands and perform various tasks. These systems are among the most common examples of artificial communication (Jurafsky & Martin, 2023, p. 315). In addition, chatbots are widely used in customer service. These bots are designed to answer customer inquiries, provide information, and even perform certain tasks. For example, banks and e-commerce sites frequently utilize chatbots to automate customer service (Shum et al., 2018).

Returning to the important role of artificial communication in education, educational platforms that offer personalized learning experiences to students support teachers and make the learning process more effective. For example, AI-based learning management systems monitor students' performance and provide customized content tailored to their needs (Holmes et al., 2019, pp. 9–10, 31–32). At the same time, AI provides significant support to teachers. With AI-powered tools, teachers can closely monitor student performance and intervene when necessary. These technologies enable teachers to track students' progress and create more effective lesson plans tailored to individual needs. Additionally, the data provided by AI helps teachers identify which subjects students are struggling with or where their talents lie (Chaudhry & Kazim, 2022, p. 159). Additionally, AI plays an important role in improving the quality and accessibility of educational materials. AI-powered tools provide students with access to a wide range of high-quality educational resources. This capability, especially through the advanced translation and customization features of AI systems, helps overcome language and cultural barriers, allowing students to access resources written in languages they do not know. Moreover, language learning applications offer activities that facilitate students' access to educational materials in different languages, helping them improve their language skills (Holmes et al., 2019, pp. 28–29).

AI is also used in learning analytics and feedback processes. Learning analytics employs data mining and machine learning techniques to monitor and analyze students' learning processes. Through this, it can be determined which subjects students are struggling with or excelling in. Teachers can use this data to provide more effective feedback to students and optimize their learning processes (Chaudhry & Kazim, 2022, pp. 158–159). Learning analytics is carried out by collectively analyzing student data and integrating the insights gained from these analyses into teaching processes. This process allows teachers to better understand student performance and develop customized educational strategies based on individual learning needs. Learning analytics can also help assess the effectiveness of educational materials and methods. Data analysis can be used to determine how students respond to course materials and which teaching methods are more effective (Paredes & Chung, 2012). These data allow for the continuous improvement of teaching materials and strategies. Moreover, learning analytics helps students better understand their own learning processes and actively engage in them. By monitoring their performance and receiving feedback, students gain a clearer understanding of which areas they need to focus on to achieve their learning goals (Arnold & Pistilli, 2012).

Learning analytics also has the potential to enhance equal opportunities in education. The analysis of student data allows for the identification of disadvantaged student groups and the development of specialized support programs for these students. This contributes to ensuring fairness in education and helps all students gain access to equal learning opportunities (Pardo & Siemens, 2014). In conclusion, AI-powered learning analytics is a crucial tool for improving feedback and learning processes in education, thereby enhancing student success and promoting equal opportunities in education. The effective use of these technologies enables both teachers and students to manage educational processes more efficiently and effectively.

Similarly, AI is also used in the healthcare sector. AI-based systems support healthcare professionals in areas such as patient monitoring, diagnosis, and treatment planning. For example, healthcare chatbots can analyze patients' symptoms, offer possible diagnoses, and provide recommendations for doctor visits (Quazi et al., 2022, p. 212). These systems can continuously monitor patients' health conditions and detect emergencies early, increasing the chances of timely intervention. Additionally, AI-powered image processing techniques are used in fields such as radiology and pathology for analyzing medical images, enabling earlier and more accurate diagnoses of diseases (Esteva et al., 2017). Another significant contribution of AI is the creation of personalized treatment plans. By considering patients' genetic information and medical histories, it becomes possible to determine the most effective treatment methods. This is especially important in cancer treatment, as each patient's response to therapy can vary (Gambardella et al., 2020, p. 2). At the same time, AI-based systems are reducing healthcare costs and increasing access to medical services. For example, through telemedicine applications, patients can receive remote medical consultations without the need to physically visit healthcare facilities (Reddy et al., 2019, p. 24). These innovations enable healthcare services to reach a broader audience and be delivered more effectively.

Social Media and Communication Platforms: Artificial communication is used to enhance content management and user interaction on social media platforms. For example, Facebook and Twitter analyze user interactions to provide personalized content recommendations and filter out spam content (Gillespie, 2018, p. 97). These platforms use data obtained from user interactions such as likes, shares, and comments to highlight content tailored to their interests. While this enriches the user experience, it also allows the platforms to keep their users engaged for longer periods (Jha & Verma,

2024, p. 181). Additionally, AI algorithms are used to detect and limit the spread of misinformation on social media platforms. For example, Facebook uses AI-powered fact-checking systems to prevent the dissemination of false information. (Bontridder & Poullet, 2021, pp. 10–11).

Another significant contribution of AI is the development of algorithms designed to enhance user safety and detect harmful content, such as hate speech. These algorithms help identify harmful content on platforms and prevent its spread (Gillespie, 2020, p. 2). Additionally, artificial communication helps brands and businesses manage customer interactions on social media. Chatbots and automated response systems quickly answer customer inquiries, thereby increasing customer satisfaction (Baethge et al., 2016, p. 277). These systems enable businesses to maintain customer service 24/7 and increase operational efficiency. In this sense, artificial communication is widely used on social media platforms to improve user experience, enhance security, and boost business effectiveness.

1

Foundations of Artificial Intelligence

Machine Learning and Deep Learning
Basic Concepts

Machine learning (ML) is a branch of AI that enables computers to learn through data and experiences. In this process, algorithms are used to analyze large datasets and make inferences from this data. ML can be categorized into different types, including supervised learning, unsupervised learning, and reinforcement learning. Supervised learning relies on labeled datasets to train models, whereas unsupervised learning utilizes data without labels. In contrast, reinforcement learning enhances decision-making by employing reward and penalty systems (Sarker, 2021, p. 13). Deep learning (DL) is a subfield of ML that uses multi-layered neural networks to automatically learn features from data, achieving successful results, particularly in complex tasks such as image, speech, and text recognition. Specialized neural network types, such as convolutional neural networks (CNNs) and recurrent neural networks (RNNs), are optimized for specific types of data. CNNs are mainly used for tasks like image processing and object recognition, while RNNs are effective for sequential data, such as time-series data and language processing.

Algorithms and Techniques

Supervised learning is a ML method in which labeled data is used to train a model. Here, each input data point has a corresponding correct output (label). The model is trained on this labeled dataset, meaning that the learning process involves understanding which inputs are associated with which outputs. For example, an image recognition system is trained using labels on specific images (e.g., "cat," "dog"). The training process enables the model to learn these relationships and optimize its ability to make accurate predictions on new, unlabeled data (Russell & Norvig, 2010, p. 695).

Supervised learning is typically used in classification and regression problems. In classification problems, the model categorizes data into predefined categories. For example, an email classification system sorts emails as "spam" or "not spam" using supervised learning. In regression problems, the model predicts a continuous value of a variable. For instance, a model used to predict house prices is trained with various house features (square footage, number of rooms, etc.) and the price labels associated with these features (NYC Data Science Academy, n.d.).

Unsupervised learning works with unlabeled data and aims to discover meaningful structures and patterns from that data. In this method, the correct answers for the input data are not provided; the model independently discovers relationships and groupings within the data. Unsupervised learning is used to identify the natural clusters or structure of the data (Murphy, 2012, pp. 9–13).

One of the common applications of unsupervised learning is clustering algorithms. Clustering divides data points into groups based on similar characteristics. For example, a clustering algorithm used for customer segmentation can group customers based on their purchasing habits. This grouping can be used to optimize marketing strategies. Another unsupervised learning technique is dimensionality reduction methods. Dimensionality reduction makes data more understandable by eliminating unnecessary or low-information features in high-dimensional datasets. Principal component analysis (PCA) and t-distributed stochastic neighbor embedding (t-SNE) are examples of these techniques. These methods are commonly used in fields such as data visualization and feature selection (van der Maaten & Hinton, 2008).

Supervised learning and unsupervised learning are complementary methods in data analytics and modeling processes. In this context, supervised learning improves model accuracy with labeled datasets, while unsupervised learning uncovers hidden structures in data, offering broader perspectives. Both methods have strengths and weaknesses in different application areas, and when used correctly, they make significant contributions to data science and ML.

DL is a subfield of ML that creates multi-layered learning models using artificial neural networks (ANNs). DL analyzes and learns from complex datasets by utilizing multi-layered neural networks that mimic the structure of the human brain. This technique has gained significant attention in recent years due to its ability to work with large datasets with high accuracy (LeCun et al., 2015).

The fundamental building block of DL is ANNs. Neural networks consist of interconnected artificial neurons, with each neuron performing a specific function. Multilayer perceptrons (MLP) are composed of multiple layers, including an input layer, hidden layers, and an output layer. The input layer receives the data, which is then processed in the hidden layers. Finally, the output layer produces the final results (Mohammadzadeh et al., 2022).

Special neural network types, such as CNNs and RNNs, are optimized for specific types of data. CNNs are particularly used for tasks like image processing and object recognition, while RNNs are effective for sequential data, such as time series and language processing. In this context, CNNs are easier to train because they have far fewer connections and parameters, although their theoretical best performance is likely only slightly worse (Krizhevsky et al., 2012, p. 1097).

ML and DL are key fields that form the foundational pillars of modern data science and AI. ML consists of algorithms that enable computers to learn from experience and make predictions about future events or data. In this process, different methods such as supervised learning, unsupervised learning, and reinforcement learning are employed. In summary, supervised learning trains models using labeled datasets to produce accurate results, while unsupervised learning focuses on discovering meaningful patterns and structures from unlabeled datasets. Additionally, reinforcement learning optimizes decision-making processes through reward and punishment mechanisms.

In light of this information, when we summarize and integrate these concepts with examples, it can be said that supervised learning is typically used in classification and regression problems, enabling the model to learn specific patterns in training data and make accurate predictions on new, unlabeled data. Classification involves sorting data into predefined categories, while regression predicts continuous values of a given variable. Unsupervised learning, on the other hand, reveals hidden structures and groups in data using methods such as clustering and dimensionality reduction. Clustering facilitates data analysis by grouping data points with similar characteristics, while dimensionality reduction makes high-dimensional datasets more interpretable and manageable. In conclusion, ML and DL are crucial for providing complementary methods in data analytics and modeling processes. The integration of these algorithms and methods allows for successful results in analyzing complex datasets and producing high-performance outcomes. The proper and, most importantly, ethical use of these technologies offers significant advantages across many areas of modern society.

Types of Artificial Intelligence

AI has become a part of the international agenda more rapidly and significantly than many previous technologies. For instance, ChatGPT first emerged in June 2018 with a relatively simple structure. However, over time, it has evolved into a significant technology accessible to everyone. When this technology was initially introduced, it was generally met without much concern or anxiety. However, over time, discussions have arisen about its potential negative impacts, such as unemployment, inefficiency, and lack of emotional sensitivity. The dangers it could pose, particularly if it falls into the hands of malicious individuals, have received widespread media coverage and occupied public discourse.

As AI technologies continue to spread rapidly, they have led to many developments, both positive and negative. With the increasing number of users, AI's capacity and speed have also grown, which has deepened and amplified the fear associated with this technology. For the first time in history, humanity is faced with a technology that can take action without human approval or control. This situation has led to AI being perceived not as a revolutionary innovation but as a source of discomfort and uncertainty.

To understand why humans fear and are affected by AI, it is essential to first examine in detail what AI is, its types, and its principles. However, even with a brief analysis, it can be concluded that new media tools such as electricity, computers, the internet, and mobile devices cannot act without human intervention, decision, and influence. In contrast, AI represents the first instance in technological history where a technology can exhibit an independent will to take action. This contributes to its image as a source of unease and uncertainty.

If we think of AI as a child learning to walk, we must consider that children at that age acquire their first education and knowledge from their families, schools, and environments. AI should be evaluated in the same manner. Just as a person needs education and certain principles to become a beneficial member of society, it is crucial to view AI as an immature child. The reality that AI could approach the world's problems in an unregulated, uncontrolled, incompetent, and ignorant manner must not be overlooked. These precautions will help prevent or minimize the potential harms that a technology with undefined boundaries may cause.

AI refers to a broadly defined structure that encompasses a diverse range of technologies. In this context, it is categorized into three main types: Narrow AI, General AI, and Super AI. Narrow AI refers to systems designed to perform specific tasks and is the most commonly used type of AI today. For instance, voice assistants and recommendation systems are examples of Narrow AI applications (Interaction Design Foundation - IxDF, 2023). General AI refers to systems capable of performing a wide range of tasks at the level of human intelligence. This type of AI possesses the ability to think like a human and make effective decisions across various domains (Built In, n.d.). Super AI, on the other hand, refers to systems that surpass human intelligence and possess the ability to learn autonomously. This type of AI is still in the theoretical stage and is considered one of the potential developments for the future (Harvard Science Review, n.d.).

This classification is fundamentally important for understanding how far the evolution of AI can progress and the potential impacts of these technologies. While Narrow AI focuses on facilitating daily life, General AI and Super AI signal future technological revolutions with their broader and more in-depth decision-making capabilities. These advancements provide significant insights into how technology can transform human life, while also introducing new debates on ethics and control. In this context, the future of AI emerges as both an exciting and a carefully considered topic. The efficiency of Narrow AI in specific tasks and the broad application potential of General AI increase trust in technology, whereas the risks and ethical concerns posed by Super AI must be seriously evaluated. However, these evaluations should aim to optimize the impact of AI on society and minimize potential negative consequences. Therefore, understanding the different levels of AI is crucial for keeping track of technological progress and predicting the direction of future developments. The fact that this is a complex process requiring a balance between scientific and ethical considerations remains a reality we continue to face.

Narrow Artificial Intelligence

Narrow AI is a form of intelligence that can perform a specific task and is limited to that task. This particular form of AI excels in areas such as visual recognition, utilizing predefined language structures, processing voice commands, and matching repetitive patterns. Its success in automation, data analysis, and various recommendation systems provides significant advantages by reducing workload, costs, and errors. Applications like Siri, Alexa, Google Translate, and Bard are widespread examples of this type of AI. Major companies such as Meta, Amazon, eBay, and Twitter, as well as important firms outside the U.S. like Huawei, Alibaba, JD.com, and Rakuten, actively use this technology.

The advantages of Narrow AI include factors such as reducing human errors, increasing efficiency, and lowering costs. In this context, Narrow AI has the potential to surpass human intelligence in specific tasks, presenting a significant opportunity in terms of economic efficiency (Nancholas, 2023). This situation, as noted by Goodfellow, Bengio, and Courville, represents a process that is evolving through the advancements in AI made possible by DL algorithms (2016, p. 486).

However, the disadvantages of Narrow AI should not be overlooked. Concerns regarding unemployment and security, the need for continuous oversight of current automation, and the presence of software errors indicate that this technology is still in its developmental stage. Brynjolfsson and McAfee have discussed the impact of automation on the labor market, stating that AI and automation technologies have the potential to entirely eliminate certain types of jobs. They argue that when such technologies render a specific job, or even an entire category of skills, obsolete, workers will need to develop new skills and find new employment (2014, p. 133). For instance, Narrow AI systems used in the automotive industry operate on production lines with high accuracy, minimizing human errors. However, this also leads to job losses for assembly line workers. Frey and Osborne highlight the rise of automation, predicting that it will cause a large-scale restructuring in the labor market (2017, p. 254–255) emphasizing that this transformation is inevitable.

In conclusion, while the advantages of Narrow AI are of great importance in terms of optimizing business processes and reducing costs, its disadvantages must also be considered. The risks of unemployment and the security threats posed by software errors highlight the need for this technology to be developed in a sustainable and reliable manner. Additionally, the need to verify the information obtained through this technology presents another significant issue. One of the biggest shortcomings of AI, particularly in language-based tools, is the generation of fabricated information when accurate data is not accessible. Russell and Norvig foresee that Narrow AI will expand into broader application areas in the future, stating that advancing with consideration of the ethical and social implications of this technology is critical for the responsible development of AI (2010, p. 1020).

General Artificial Intelligence

General AI refers to systems that can think and learn like the human brain. These systems are capable of performing tasks independently, making decisions, and taking action. While this technology can exhibit human-like behaviors in all areas, it is still in the developmental stage, and it is certain that this phase will not be completed in the near future. This situation also brings with it various discussions and concerns about AI. One of the greatest fears is the loss of control over AI. Numerous urban legends, such as the story of Facebook engineers shutting down an AI system after losing control of it, have spread rapidly on social media. Such unverified content often gains traction among large audiences, becoming widely circulated due to its intriguing nature.

General AI has the potential to profoundly transform fields such as medicine, architecture, and economics. However, the unrestricted use of this technology could lead to uncertain outcomes in terms of ethical principles and security. Therefore, specific guidelines must be established during the development of General AI, and adherence to these principles should be ensured as much as possible. Limiting a technology that possesses human-like thinking and decision-making capabilities can help minimize potential harms.

Research on the ethical and societal impacts of AI indicates that this technology can give rise to serious ethical issues related to human rights, justice, autonomy, and accountability. Particularly, the use of AI in the healthcare sector raises significant concerns regarding data privacy and security (Li et al., 2023). Additionally, the use of AI in the business world could lead to unemployment and job displacement, while also requiring workforce adaptation (Trotta et al., 2023).

The uncontrolled use of General AI could lead to broad-ranging issues that are not only technological but also social and ethical in nature. Therefore, the development and application of AI must adopt ethical principles, and these principles should be supported through international collaboration. In the future, comprehensive policies and regulations should be established to ensure the safe and ethical use of AI (Floridi & Cowls, 2019).

When considering the development and potential impacts of General AI, several important conclusions can be drawn. First, the ability of General AI to think and make decisions like humans highlights its potential to bring about profound changes in many fields. It is particularly anticipated to introduce major innovations in areas such as medicine, architecture, and economics. However, with the development of this technology, there is also a risk of serious ethical and societal issues emerging.

At this point, the fear of losing control over AI and the urban legends circulating on this topic are causing significant concern in society. The rapid spread of such rumors can undermine public trust in technology and lead to unnecessary fears. Therefore, during the development of AI, it is crucial to prevent the dissemination of unverified information and to raise public awareness.

The necessity of adopting specific ethical principles during the development of AI and adhering to these principles plays a critical role in minimizing the potential harms of the technology. Ethical principles should serve as a guide to ensure the safe and responsible use of AI. Supporting these principles through international cooperation will enable AI to be used more safely and ethically on a global scale. Moreover, research on the development and applications of AI shows that the technology can raise significant issues related to fundamental ethical concerns such as human rights, justice, autonomy, and accountability. Therefore, comprehensive policies and regulations should be established, taking into account the societal impacts of AI. These regulations should be designed to minimize the negative effects of AI and ensure that the technology is used for the benefit of humanity. Additionally, while the use of AI in the business world could lead to unemployment and job displacement, it also necessitates that the workforce adapts to this new situation. In this context, both the potential benefits and risks of General AI must be considered, and ethical and safety principles should be meticulously followed during the development of this technology. This will not only ensure the healthy progression of the technology but also foster society's trust in this advancement.

Super Artificial Intelligence

Super AI refers to systems that are anticipated to possess an intelligence far beyond human comprehension and may also include emotional components. This type of AI is expected to have the ability to think, learn, and evaluate at a level and speed unreachable by humans. While this could provide significant contributions to humanity, it also poses potential dangers. Super AI could accelerate technological progress to an unprecedented degree, leading to many positive effects for human history, but if used for harmful purposes or if control is lost, it could result in negative outcomes. It is predicted that the development process of Super AI will be completed in a shorter time compared to other new media technologies. This makes Super AI an important issue that requires ethical, legal, financial, and physical regulations.

The establishment of ethical principles and regulations during the development and application of Super AI is of great importance. For example, research on the potential of Super AI in the healthcare sector shows that this technology has the capacity to personalize healthcare services and increase accessibility, but it also brings ethical issues. These issues include data privacy, bias, transparency, and accountability (Lund & Ghiasi, 2024).

In addition, the use of Super AI in the business world raises significant ethical and societal questions. It is noted that while Super AI could optimize business processes and increase efficiency, it could also exacerbate issues such as unemployment and social inequality. Therefore, during the development and implementation of Super AI, policies and strategies must be developed to facilitate the adaptation of the workforce (Trotta et al., 2023).

The uncontrolled use of Super AI could lead to wide-ranging issues that are not only technological but also social and ethical in nature. Therefore, the development and application of AI must adopt ethical principles, and these principles should be supported through international cooperation. In the future, comprehensive policies and regulations must be established to ensure the safe and ethical use of AI. Adherence to ethical principles will not only ensure that AI benefits humanity but also minimize its potential harms. In this context, the development and use of AI have the potential to produce both positive and negative impacts on society; it could alleviate or exacerbate existing inequalities, solve old problems, or create new ones. Determining the preferred societal path depends not only on well-prepared regulations and common standards but also on the use of an ethical framework that can guide concrete actions (Floridi & Cowls, 2019, p. 10).

The concept of Super AI represents a technological advancement that holds great opportunities for humanity, while also posing significant risks. This type of intelligence offers learning and thinking abilities that surpass human capacity. In addition to the potential benefits of Super AI in healthcare and business, warnings must also be considered regarding the important issues this technology could raise, such as data privacy, bias, and unemployment. Furthermore, it is noted that if this technology is used without control, it could lead to complex problems not only on an individual level but also at societal and global levels.

In this context, the importance of ethical principles and international cooperation in the development and application of Super AI is emphasized. Ethical frameworks aim to ensure that such technology benefits society rather than causing harm. However, to achieve these goals, theoretical ethical principles alone are not sufficient; concrete global policies and regulations must also be established. The effects of Super AI may either exacerbate or alleviate existing societal inequalities, depending on how it is used and what regulations are applied.

In conclusion, a careful balance must be struck between the potential benefits and risks of Super AI. It is clear that if the technology is used uncontrollably and in violation of ethical principles, it poses serious dangers to humanity. Therefore, determining the preferred societal path must be supported by concrete actions grounded in ethical principles. Managing the opportunities that Super AI presents in a safe and responsible manner emerges as both a necessity and a responsibility for humanity.

Natural Language Processing (NLP)
Definition and Importance
NLP is a field of AI that develops the ability of computers to understand, process, and generate human language. NLP sits at the intersection of linguistics, computer science, and AI, enabling machines to comprehend, interpret, and manipulate natural human language. The importance of this technology lies in its ability to facilitate human-machine interaction using language and to overcome language barriers, thereby easing global communication. In this context, one of the most comprehensive works on NLP, *Speech and Language Processing*, provides valuable insights. This work introduces the fundamental algorithmic tools that constitute modern neural language models and form the core of end-to-end NLP systems. The book covers useful algorithms such as tokenization and preprocessing, followed by tasks like calculating edit distance, and then delves into classification, logistic regression, ANNs, feedforward networks, recurrent networks, and finally transformers. Additionally, the role of embeddings as word meaning models is explored, making this study a highly valuable resource for understanding NLP (Jurafsky & Martin, 2023, p. 1).

NLP, by its nature, utilizes advanced algorithms and models to address the complexity and ambiguity of language, serving a wide range of applications such as text classification, sentiment analysis, translation, and speech recognition. In this regard, machine translation, which involves the automatic translation of text or speech from one language to another, is one of the most important applications of NLP (Manning & Schütze, 1999, p. 463).

NLP emerges as a revolutionary field in enabling computers to understand and process human language. As mentioned, the use of advanced algorithms and models to address the complexity and ambiguity of language highlights the importance and versatility of NLP. In this context, it is worth revisiting *Speech and Language Processing* by Jurafsky and Martin. This work, which stands out as a significant resource covering the fundamentals and modern applications of NLP, provides an in-depth guide to NLP studies. It spans a wide range of topics, from essential steps such as tokenization and preprocessing to more advanced techniques like classification and ANNs.

The role of NLP in our daily lives is increasingly growing. Applications such as machine translation, in particular, are breaking down language barriers, facilitating global communication, and enabling people from different languages to communicate more effectively. As emphasized in the work of Manning and Schütze, machine translation is one of the most important application areas of NLP. It refers to the process of automatically translating a text or speech from one language to another, and this process plays a major role in accelerating global interactions.

The possibilities offered by NLP are not limited to translation or text processing; it also holds significant potential in areas such as sentiment analysis, chatbots, speech recognition, and many more. These technologies are revolutionizing fields ranging from customer service to the healthcare sector. The future development of NLP will open doors to more natural and seamless human-machine interactions, further deepening the impact of this technology. Therefore, studies on NLP will continue to have a significant influence on society as a whole, beyond just being an academic interest.

Key Techniques and Methods

The key techniques and methods used in the field of NLP involve various algorithms and models designed to analyze and understand the structural features of language. Among these methods, statistical models, DL techniques, and ML algorithms stand out. For instance, n-gram models are employed to predict the sequence of words in a language, while ANNs are used to comprehend more complex linguistic structures. (Jurafsky & Martin, 2023).

DL techniques include models such as CNNs and RNNs, which are particularly effective in extracting deeper semantic meanings of language. Transformer models, and especially large language models like Generative Pretrained Transformer (GPT)-3, are among the revolutionary methods in NLP in recent years (Vaswani et al., 2017).

These key techniques and methods play a crucial role in enhancing the performance of NLP applications. Statistical approaches, such as n-gram models, are particularly effective in analyzing the surface structures and word sequences of language, whereas DL methods excel in extracting more complex and contextual meanings. Forms of DL, such as CNNs and RNNs, are utilized to model the structure of language and its temporal dependencies.

In particular, RNNs have shown great success in capturing temporal dependencies in language and modeling the sequential features of language.

Transformer models represent a groundbreaking development in NLP. Introduced by Vaswani et al. in 2017, this model produces more accurate and contextual meanings by focusing on different parts of a sentence through the attention mechanism. These models better capture the complex structure of language and long-range dependencies, demonstrating high performance in tasks such as translation, language modeling, text generation, and various other NLP tasks. Particularly, large language models like GPT-3, with their massive architectures containing billions of parameters, offer human-like capabilities in language generation and understanding. These models have achieved impressive results in tasks such as text completion, summarization, translation, and question answering.

Another important technique used in NLP is word embeddings. This method aims to capture semantic similarities between words by representing them as continuous vectors. Techniques such as Word2Vec, GloVe, and more recent models like Bidirectional Encoder Representations from Transformers (BERT) and GPT play a crucial role in modeling the semantic structure of language by better understanding the context of words. In particular, BERT, as a bidirectional language model, can extract richer and deeper meanings by considering the context on both sides of a sentence (Devlin et al., 2019).

These advanced techniques and methods have paved the way for significant advancements in many areas of NLP. For instance, in applications such as text classification, sentiment analysis, machine translation, and natural language generation, the use of these models and algorithms enables the production of more accurate and human-like results. The progress of NLP in these areas is not only limited to academic research but also manifests in practical applications that impact our daily lives.

In this context, the key techniques and methods used in NLP continue to provide advanced algorithms and models that successfully analyze both the surface-level and deep meanings of language, addressing its complex structure. Future developments in these technologies will further enhance machines' capacity to understand and generate human language, accelerating the pace of innovation in this field.

Current Applications and Research

NLP is currently being widely applied across various fields. For instance, speech recognition systems convert users' voice commands into text and are commonly used by voice assistants such as Siri and Google Assistant (Xiong et al., 2017). Translation services, particularly tools like Google Translate, facilitate international communication by automating interlingual translation (Wu et al., 2016). Sentiment analysis is used to evaluate users' emotional responses in areas such as social media analysis and customer feedback. Additionally, NLP enhances user experience in human-computer interaction applications, such as chatbots and virtual assistants (Radziwill & Benton, 2017).

In the field of research, advancements in NLP have led to the development of more sophisticated approaches to language modeling. Notably, models such as GPT and BERT have shifted the paradigm of how large language models are trained and applied (Devlin et al., 2019). These models are trained on large datasets to better understand the context of language and make accurate predictions, producing highly effective results in various language tasks.

These impressive advancements in NLP have led to transformative changes across many industries. For example, in the healthcare sector, the use of NLP has become widespread in processes such as processing electronic health records and automatically coding doctors' notes (Patra et al., 2021). These and similar advanced applications enable faster and more accurate analysis of medical data, improving patient care.

In the field of financial services, NLP algorithms are used to analyze market sentiment, optimize investment strategies, and detect fraud (Chancellor et al., 2019). In particular, data obtained from news articles and social media posts is regarded as a significant source for predicting market trends.

In the education sector, NLP-based tools offer personalized learning experiences to students and provide feedback to teachers, making educational processes more effective. Automated assessment systems and language learning applications track students' progress and provide materials tailored to their individual needs. In this context, education reforms in recent years have been developed worldwide with the aim of preparing citizens for the challenges of globalization (Liang et al., 2009, p. 69).

The future potential of NLP is developing in a way that will provide more human-like interactions, while also considering ethical and security concerns.

In particular, ethical AI practices, issues of data privacy, and algorithmic fairness stand out as important areas for evaluating the societal impacts of NLP. In this regard, various costs and risks are being identified alongside the increasing race to develop larger language models. These include environmental costs, financial constraints, and the significant risks of harm resulting from the misinterpretation of language model outputs (Bender et al., 2021, p. 619). Therefore, the responsible and fair development of NLP plays a critical role in ensuring the long-term sustainability of this technology.

Computer Vision
Image Processing and Analysis

Computer vision is a branch of AI that aims to enable computers to extract meaningful information from digital images or videos. Image processing is a fundamental step that involves operations such as enhancing, filtering, and extracting features from digital images. The image processing process allows computers to make sense of raw pixel data and derive meaningful insights from it (Gonzalez & Woods, 2002, p. 1). In this process, a wide range of techniques is used, from pixel-based operations to high-level object recognition tasks. For example, techniques such as edge detection, histogram equalization, and morphological operations are commonly used for enhancing and analyzing images (Szeliski, 2022, p. 109).

Image analysis is a process that follows image processing, focusing on interpreting objects and scenes within images. This process is used for computers to recognize, categorize, and analyze specific objects in images. Techniques used in image analysis often involve complex ML methods, such as ANNs and DL models. In particular, CNNs are highly effective in image analysis, demonstrating strong performance in tasks such as feature extraction and object recognition (LeCun et al., 2015, p. 436).

Image processing and analysis play a critical role in many areas of modern technology. The process of extracting meaningful information from digital images is not limited to enhancing images and improving visual quality, but also offers a wide range of applications through the advanced analysis of this data. In this context, making sense of raw pixel data is a fundamental process for computer vision systems. It is clear that this process is necessary for performing more complex tasks such as object recognition and scene analysis.

Facial Recognition and Object Detection

Facial recognition is one of the most well-known and widely used applications of computer vision. This technology is used to identify and recognize faces in digital images. Facial recognition systems are commonly used in areas such as security, identity verification, and enhancing user experience. The facial recognition process employs various algorithms to extract the contours and specific features of a face, then compares these features with previously stored databases to identify individuals (Tolba et al., 2005, p. 88). Deep learning-based methods, particularly CNNs, work with high accuracy in facial recognition tasks by automatically learning facial features without the need for manual feature extraction (Bansal et al., 2021, pp. 6–29).

Object detection is another important application area of computer vision. Object detection is the process of identifying specific objects in an image and determining their locations. This technology is commonly used in security systems, autonomous vehicles, and robotic systems. Object detection algorithms are typically based on techniques such as sliding window, regional proposal networks (R-CNNs), and CNNs (Girshick et al., 2014, p. 1). Notably, models such as YOLO (You Only Look Once) and SSD (Single Shot Multibox Detector) stand out in object detection for providing high speed and accuracy. (Redmon et al., 2016).

Facial recognition and object detection, as two fundamental components of computer vision technologies, have a wide range of applications in various fields of the modern world. These technologies are not only limited to security and identity verification but also offer innovative solutions in industries such as marketing, healthcare, automotive, retail, and many more. Facial recognition technology, in particular, is frequently encountered in everyday applications like unlocking smartphones. For example, Apple's Face ID technology uses a deep learning-based method to secure the device by recognizing the user's face. This technology creates a three-dimensional model of the face, going beyond two-dimensional images and enabling recognition from different angles.

In terms of security, facial recognition technology is used in airports, public areas, and various security checkpoints. For instance, many airports use facial recognition systems for passenger identification, speeding up passport control processes. Additionally, facial recognition technology plays an active role in locating missing persons or tracking suspects. In countries like China, large-scale facial recognition systems deployed across cities, combined with big data analytics, have become a crucial tool in combating crime.

In addition, object detection stands out as a vital technology, particularly for autonomous vehicles. Autonomous vehicles use object detection algorithms to detect and identify other vehicles, pedestrians, traffic signs, and obstacles around them, making decisions based on this data. For example, Tesla's autonomous driving technology enhances driving safety by detecting and classifying objects around the vehicle in real time. Furthermore, object detection plays an important role in the retail sector. In-store security cameras are equipped with object detection technology to prevent theft. This technology is also used to track the inventory status of products on shelves and detect missing items. Amazon's cashierless store concept, Amazon Go, has revolutionized the shopping experience by combining object detection and AI technologies.

Facial recognition and object detection technologies will continue to evolve alongside advances in AI and DL algorithms. For example, in the future, these technologies could be more widely used in the healthcare sector for early diagnosis and treatment processes. Facial recognition could be used to detect symptoms of neurological diseases, while object detection could enable surgical robots to work more precisely and effectively. In this context, facial recognition and object detection technologies have a transformative impact not only in security and automation but also in many different aspects of life. The correct and ethical use of these technologies will play a critical role in translating future developments into societal benefit.

Application Examples

Computer vision has become a technology of critical importance for many industries and fields today. This technology is widely used in sectors such as healthcare, agriculture, security, automotive, and entertainment. For instance, in the medical field, computer vision techniques are employed in the analysis of medical images, early diagnosis of diseases, and monitoring of treatment processes. Specifically, the automatic analysis of medical images like MRI and CT scans provides an important tool to assist doctors in the diagnosis process (Litjens et al., 2017).

In the agricultural sector, computer vision is used in applications such as monitoring crop health, detecting pests, and yield prediction. Images obtained through drones and satellites are used for the analysis and optimization of agricultural fields, providing farmers with information to enhance productivity (Kamilaris & Prenafeta-Boldú, 2018, p. 2).

In the automotive sector, computer vision plays a vital role in the development of autonomous vehicles. This technology enables vehicles to perceive their surroundings, recognize objects, and move safely. The integration of LiDAR, radar, and camera data used in autonomous driving systems allows vehicles to make real-time decisions and adapt to their environment (Chen et al., 2015).

In the entertainment sector, computer vision technologies are used in augmented reality (AR) and virtual reality (VR) applications. These technologies enable users to interact between physical and digital worlds, offering new experiences in areas such as gaming, education, and social media. Technological demands for AR, in particular, are much higher than those for virtual environments or VR, which is why the maturation of the AR field has taken longer than that of VR (van Krevelen & Poelman, 2010, p. 2).

Computer vision technology also plays a critical role in the security and defense sectors. Features such as facial recognition, motion detection, and crowd analysis in security cameras and other surveillance systems are made possible by computer vision algorithms. These systems are used in areas such as crime prevention, intrusion detection, and rapid response in emergencies. For example, surveillance systems in large cities can analyze crowd behavior using computer vision techniques and detect potential threats in advance (Zhao et al., 2003, p. 424).

In the defense field, computer vision technologies are used in the operations of unmanned aerial vehicles (UAVs) and other autonomous systems. UAVs can carry out tasks such as detecting, tracking, and executing operations against enemy targets using computer vision algorithms, without the need for human intervention. These applications enhance the effectiveness of military operations and minimize human casualties. In this context, the wide range of law enforcement applications has become feasible after nearly 30 years of research (Zhao et al., 2003, p. 400).

Additionally, computer vision technologies are increasingly being used in the retail sector. In-store customer behavior analysis, inventory management, and product placement are optimized using computer vision techniques. For instance, tasks such as counting products on shelves, stock control, and theft prevention are automatically performed by computer vision-based systems (Abdul Hussien et al., 2021, p. 6).

Finally, computer vision technology has also become a significant tool in environmental monitoring and sustainability. This technology is used in applications such as monitoring environmental pollution, forest fires, and water resources. For instance, satellite imagery can be used to rapidly detect and respond to deforestation. Similarly, the quality and quantity of water resources can be monitored through computer vision algorithms, providing critical data for sustainability efforts (Lopatin et al., 2016).

This wide range of applications demonstrates how powerful and versatile computer vision technology is. As it is being used in more and more fields every day, this technology will continue to develop and increase its impact on all areas of life in the future.

2

Artificial Communication Tools and Technologies

Chatbots and Virtual Assistants
History and Development
The development of chatbots and virtual assistants is closely related to the evolution of AI and natural language processing (NLP) technologies. The first example of a chatbot, "ELIZA," was developed by Joseph Weizenbaum at MIT in 1966. ELIZA was a program that could analyze user inputs based on specific keywords and generate simple responses. Weizenbaum designed this program to mimic the Rogerian psychotherapy model (Weizenbaum, 1966). ELIZA's success was the first tangible step in demonstrating the potential of NLP technologies.

In the 1970s, "PARRY" was developed, following in the footsteps of ELIZA. PARRY simulated the mental state of a paranoid schizophrenic patient and had a more complex structure compared to ELIZA (Colby et al., 1975). In the 1980s and 1990s, advancements in AI and NLP enhanced the capabilities of chatbots. However, during this period, chatbots remained primarily experimental projects.

The 2000s marked the beginning of the integration of chatbots and virtual assistants into commercial applications, driven by the widespread use of the internet and the increase in data processing capacity. "ALICE" (Artificial Linguistic Internet Computer Entity) won the Loebner Prize in 2001 and became a significant milestone in the development of chatbot technologies (DevX, n.d.). However, a true turning point occurred in 2011 with Apple's introduction of the virtual assistant "Siri." Siri was the first large-scale commercial virtual assistant capable of recognizing voice commands and performing various tasks. This was followed by other virtual assistants such as Google Now, Amazon Alexa, and Microsoft Cortana.

The development of chatbots and virtual assistants has accelerated further with technologies such as machine learning (ML), big data analysis, and cloud computing. Today, these systems are used in a wide range of fields, from customer service to healthcare, education, and personal assistant services.

Working Principles

Chatbots and virtual assistants utilize various technological components to process user inputs and generate meaningful responses. The fundamental working principles of these systems encompass areas such as NLP, ML, and knowledge base management.

NLP is the process of making human language understandable to computers. NLP performs morphological, syntactic, and semantic analysis of the language to interpret the input from the user. For example, a chatbot analyzes the key words and grammatical structures in a sentence written by the user and generates an appropriate response by understanding them (Jurafsky & Martin, 2023, p. 327). NLP not only analyzes the surface-level features of language but also examines its deeper semantic layers. In this process, elements such as phrases, sentence structures, and the context within the text are considered. For example, a virtual assistant performs semantic analysis of the language to understand the underlying intent of a user's question. These analyses enable computers to interpret texts or speech more effectively and interact with users in a natural manner. Additionally, the evolving techniques of NLP offer significant contributions in areas such as sentiment analysis, machine translation, and automatic summarization.

ML enables chatbots and virtual assistants to continuously learn and improve. This technology allows systems to learn from past data and provide more accurate and personalized responses in future interactions. For example, a chatbot used on an e-commerce site can analyze users' previous shopping data and offer personalized product recommendations (Goodfellow et al., 2016, pp. 77–122). ML also enhances the language understanding capabilities of chatbots and virtual assistants, enabling them to interact with users in a more natural and meaningful way. For example, a customer support bot can better understand users' complaints and offer appropriate solutions. Additionally, ML algorithms become more sensitive in understanding users' language structures and expressions, thereby improving the quality of interactions.

ML also helps chatbots evaluate their own performance over time and reduce their error rates. These systems continuously improve their responses by analyzing feedback from users. As a result, chatbots and virtual assistants become more capable of understanding user needs and providing the most suitable services. In this context, ML enables chatbots and virtual assistants to evolve from being tools that execute simple commands to becoming sophisticated systems capable of deeper interactions with users and offering personalized services.

Knowledge Base Management is essential for chatbots and virtual assistants to provide accurate information. These systems can access large amounts of information on specific topics and use that data to answer user questions. For example, a healthcare chatbot can provide information about a symptom by utilizing a medical knowledge base. Additionally, chatbots are used in education for tasks such as answering frequently asked questions, handling administrative tasks, mentoring students, providing motivation, assessing student learning, simulations, training, and giving feedback. For instance, at Cardenal Herrera University, a chatbot serves as a personal assistant to monitor students, predict their behavior, answer administrative questions, and offer advice (Aleedy et al., 2022). In this context, knowledge base management is a critical component that directly affects the effectiveness of chatbots and virtual assistants. The quality, accuracy, and currency of the knowledge base determine the reliability of the information these systems provide to users. Knowledge base management not only involves collecting and organizing data but also continuously updating and verifying this information. This is particularly crucial in sensitive fields like medicine and law, where incorrect information can have negative consequences for users.

Chatbots and virtual assistants use the data from their knowledge bases, along with NLP techniques, to interpret users' questions and generate appropriate responses. In this process, ML algorithms and AI techniques enable better understanding of user queries and more accurate generation of responses. For example, a virtual assistant used in a law firm can provide clients with general legal information, offer updates on legal processes, or instantly respond to frequently asked questions. However, to handle more complex questions at the level of legal consultancy, these systems require deeply trained knowledge bases and advanced AI models.

In the education sector, knowledge bases are used to track student achievements, provide personalized learning materials, and support individual learning pathways. For chatbots and virtual assistants to provide accurate and timely feedback and assessments to students, access to comprehensive and up-to-date knowledge bases is essential. These systems can also support students in exam preparation, assist with assignments, and offer motivation for overall academic success. For example, in some universities, chatbots function as dynamic guides, helping students easily access course materials and achieve their learning goals. In this sense, knowledge base management plays a central role in the success of chatbots and virtual assistants. For these systems to make informed decisions and generate accurate responses, their knowledge bases must be effectively managed and continuously updated.

In the future, with advancements in AI and ML, these systems are expected to become even more capable and perform a wider range of complex tasks. Chatbots and virtual assistants operate through the integration of these key components. User input is first processed through NLP, followed by learning and personalization via ML algorithms, and finally, the necessary data is retrieved from the knowledge base to generate a response. In this sense, these processes enable chatbots to communicate with users more naturally and effectively, incorporating systems that work in an integrated manner.

Application Areas and Success Stories

Chatbots and virtual assistants are widely used in various fields today. One of the most common areas of application for these technologies is customer service. For example, Amazon Alexa is a virtual assistant that allows users to shop through voice commands, control smart home devices, and even play music (Amazon, n.d.). Amazon Alexa is a successful example developed to facilitate users' daily lives and meet their personal needs. Another common use case is Sephora's chatbot, "Sephora Virtual Artist." This chatbot allows users to obtain information about makeup products, virtually try them on, and make purchases. By uploading a photo of their face, users can experiment with different makeup products and select the one that suits them best (Sephora, n.d.).

The healthcare sector is also a field where chatbots are commonly used. Health chatbots, such as Buoy Health, offer users symptom-checking and preliminary diagnosis services. These chatbots analyze users' inputs to provide information about potential health issues and guide them to consult a doctor when necessary (Buoy Health, n.d.). Similarly, Woebot is an example of a chatbot that provides users with daily psychological support. This chatbot monitors users' emotional states by asking questions about their feelings and offers appropriate psychological interventions when needed (Woebot, n.d.).

The education sector is another area where chatbots are used to provide student guidance and educational support. Duolingo is a popular platform that integrates chatbots to enhance the language learning process. These chatbots offer personalized feedback to students while monitoring their performance, delivering individualized learning experiences (Duolingo, n.d.). Chatbots are also used to provide student support in universities. Chapman University, for instance, employs a chatbot called PantherBot to guide students on registration processes, financial aid applications, and class schedules. This chatbot offers quick and efficient responses to common student inquiries, simplifying their academic journey (Chapman University, n.d.).

Another example is the financial sector, where chatbots greatly facilitate customer service and banking transactions. Bank of America's virtual assistant, Erica, helps users check account information, pay bills, and receive financial advice (Bank of America, n.d.). Additionally, Mastercard's chatbot "Kai" assists customers with account transactions, spending analysis, and card information. Kai allows customers to access their financial information via voice commands or written messages, enhancing the overall user experience (Mastercard, n.d.). Another similar example is the Hongkong and Shanghai Banking Corporation or HSBC's virtual assistant "Amy." Amy is a chatbot that helps HSBC customers with banking transactions, account information, and credit card applications. Amy enables users to access their financial information through written messages or the mobile app and also offers advice on personal finance management. This allows customers to perform banking tasks more quickly and easily (Conversation Design Institute, n.d.).

A similar example can be found in Levi's "Virtual Stylist" chatbot. Announced by Levi Strauss & Co., Levi's Virtual Stylist is a chatbot that personalizes the shopping experience by offering accurate size and style recommendations. When selecting jeans, users provide their measurements, preferred cut, and style preferences, and the chatbot suggests the most suitable products based on this information. Levi's Virtual Stylist aims to replicate the in-store experience in a virtual setting, making online shopping more convenient while also enhancing customer satisfaction (Levi Strauss & Co., 2017). In the tourism sector, Booking.com has integrated chatbots to assist users in planning their trips. These chatbots help customers make reservations, access hotel information, and answer travel-related questions, enabling users to organize their travel plans more quickly and easily. Another significant example is KLM Royal Dutch Airlines' chatbot, "KLM Messenger." KLM Messenger assists passengers with services such as flight reservations, boarding passes, and flight updates. Passengers can interact with KLM Messenger through Facebook Messenger, WhatsApp, and other popular messaging apps to receive information about their flights and get their questions answered. This chatbot enhances the travel experience by offering personalized services and simplifying pre- and post-flight processes (KLM Royal Dutch Airlines, 2017).

In the future, the areas of application for chatbots and virtual assistants are expected to expand even further. With advancements in AI and machine learning technologies, these digital assistants will be able to better understand human behavior and offer more personalized and proactive services. For instance, in the healthcare sector, chatbots could analyze genetic data to predict individual health risks and provide personalized health recommendations accordingly. In education, virtual assistants could enhance students' academic performance by providing real-time feedback tailored to their learning style. Similarly, in the financial sector, advanced chatbots could analyze users' spending habits to offer more effective budgeting and savings recommendations. As these technologies become more integrated into daily life, interactions between humans and machines will grow more complex, and these digital assistants will deliver increasingly sophisticated services that improve users' quality of life.

Voice Assistants and Speech Recognition
Development of Voice Assistants

Voice assistants are software applications that allow users to interact with devices through voice commands. The first voice assistant technologies were developed to recognize simple voice commands and perform basic tasks. The Audrey system, developed by Bell Laboratories in the late 1960s, is considered one of the first automatic speech recognition (ASR) systems, and it could only recognize ten digits (Liang, 2024, p. 180). In subsequent years, significant advancements were made in speech recognition technology with the development of computing power and algorithms. The launch of Apple's

Siri in 2011 brought voice assistants to a wider audience, allowing users to perform a variety of tasks in their daily lives through voice commands.

Over time, voice assistants have evolved beyond executing simple commands, becoming equipped with NLP capabilities. As a result, voice assistants can now understand more complex interactions, process contextual information, and provide responses that are more tailored to users' needs. For example, voice assistants like Google Assistant, Amazon Alexa, and Microsoft Cortana assist users with tasks such as playing music, checking the weather, managing calendars, and searching the web. These systems continue to improve over time through continuous learning and user feedback, offering increasingly better services to users.

Speech Recognition Technologies

Speech recognition technologies consist of algorithms and models used to convert voice commands into digital text. These technologies comprise various components such as acoustic modeling, language modeling, and speech processing. Acoustic modeling analyzes the acoustic features in the speech signal and associates these features with phonemes or other linguistic units. Language modeling, on the other hand, understands the linguistic structure of speech and determines the probabilistic relationships between words and sentences (Jelinek, 1997, p. 57).

Traditional speech recognition systems are built on statistical approaches such as hidden Markov models (HMMs) and Gaussian mixture models (GMMs) (El-emary et al., 2011). These models have been used to understand and recognize the probabilistic features of speech. However, with the advancements in deep learning techniques in recent years, significant progress has been made in speech recognition systems. Specifically, models such as deep neural networks (DNNs) and convolutional neural networks (CNNs) have enabled higher accuracy rates in speech recognition tasks (Hinton et al., 2012, p. 83). Additionally, recurrent neural networks (RNNs) and long short-term memory (LSTM) models have led to revolutionary advancements in speech recognition, thanks to their ability to process time-series data (Graves et al., 2013).

Speech recognition technologies are widely used in many fields today. In addition to devices such as smartphones, tablets, and computers, these technologies have been integrated into various areas, including the automotive industry, healthcare sector, and home automation. For example, in the automotive industry, voice command systems have been developed to allow drivers to control their vehicles without using their hands (Wu et al., 2022). In the healthcare sector, speech recognition technologies are used to enable doctors to record patient information through voice commands and to allow patients to monitor their health conditions (Kumar, 2024, p. 7). The use of these technologies in the healthcare sector allows doctors to maintain patient records more quickly and accurately, and it also plays a crucial role in helping patients monitor their health conditions. In particular, the ability of RNNs and LSTM models to process time-series data has brought revolutionary innovations to this field. These advancements reduce the workload of doctors while enabling patients to easily track their health status.

In the future, it is possible to foresee that these technologies will continue to evolve and become more widespread, not only in the healthcare sector but also in industries such as automotive and home automation. In the automotive industry, the hands-free capability provided by voice command systems enhances safety, while in home automation, the ability for users to control their homes with voice commands represents a significant development that simplifies daily life. In this context, it is clear that speech recognition technologies will continue to contribute to a smarter and more integrated world by improving the user experience.

Applications and Usage Scenarios

Voice assistants and speech recognition technologies have a wide range of applications, from daily life to professional settings. These technologies simplify tasks for users in various fields such as home automation, education, healthcare, and automotive. For example, in the field of home automation, platforms like Amazon Alexa and Google Assistant allow users to control their home devices through voice commands. As a result, users can turn lights on and off, adjust thermostats, and manage home security systems. The widespread use of voice assistants in areas beyond home automation demonstrates how prevalent and effective this technology has become.

In the field of education, voice assistants offer significant conveniences to students. They can guide students during the study process, provide quick access to information, and support language learning. Additionally, they offer great benefits for individuals with disabilities by making digital devices more accessible, thereby promoting equal opportunities in education. The accessibility features provided by voice assistants make the digital world more reachable, particularly for disabled individuals. The potential offered by these technologies for education and individuals with special needs has become an integral part of innovative approaches in education.

In the healthcare sector, voice assistants and speech recognition technologies reduce doctors' workloads and enable faster and more accurate patient record-keeping. Additionally, these technologies allow elderly and disabled individuals to monitor their health and seek help when necessary. Platforms like Amazon Alexa support healthcare processes by reminding users to take their medications on time. These features provide significant conveniences for both professionals and patients in the healthcare sector (Zhou et al., 2019). These advancements in the healthcare field demonstrate that technology has become an element that enhances quality of life.

In the automotive industry, voice command systems provide great convenience to drivers. Drivers can control their vehicles through voice commands and perform tasks such as navigation, playing music, and making phone calls without taking their hands off the wheel. Leading automakers such as Tesla, BMW, and Mercedes-Benz have integrated these systems into their vehicles, enhancing the user experience. These integrations not only improve driving safety but also offer drivers a more comfortable driving experience (Tepe, 2020). The adoption of voice command systems in the automotive industry demonstrates the contributions of technology to the driving experience.

In this context, voice assistants and speech recognition technologies are rapidly developing and expanding fields. Today, these technologies have become widespread in many areas, from home automation to healthcare, automotive to education. These technologies enable users to interact with digital devices in a more natural and efficient manner, and they are expected to advance further in the future, with new use cases emerging. This will create a technology ecosystem that enhances user experience and integrates into every aspect of life. Future advancements will lead to the use of these technologies in even more areas, offering new opportunities.

Social Media and Artificial Communication

Social media can be defined as digital platforms that have significantly transformed communication between individuals and communities. These platforms provide interactive spaces that allow people to exchange information, create, and share content. Unlike traditional media tools, social media enables two-way communication among users, and this interaction has been further enhanced by AI technologies.

AI is widely used on social media to enhance user experiences, optimize content management, and analyze interactions. AI-based algorithms provide content recommendations on social media platforms, filter spam content, and create personalized experiences by analyzing user behavior. For example, platforms like Facebook and Instagram use ML algorithms to offer new content recommendations based on the content users have previously liked or commented on (Sadiku et al., 2021, p. 16).

The impact of AI on social media is not limited to content recommendations and personalized experiences. These technologies are also used to ensure user safety, promote the healthy growth of online communities, and minimize negative behavior in digital environments. One of the key roles of AI on social media platforms is the detection of harmful content and preventing its spread.

AI can analyze user behavior to detect issues such as online harassment, hate speech, and misinformation. For example, platforms like Twitter and YouTube use NLP techniques to identify harmful content and take proactive measures to prevent its spread. These algorithms scan large datasets, allowing potentially harmful content to be quickly identified and removed.

Additionally, another significant impact of AI on social media is its ability to deeply analyze user interactions, determining which content attracts the most interest and how these interactions can be improved. These analyses help platforms provide more engaging and relevant content to their users. For instance, LinkedIn analyzes how users respond to job postings and, based on this data, offers better job recommendations. Such analyses increase the value that platforms offer to their users, encouraging them to stay on the platform for longer periods.

The impact of AI on social media is not limited to content management and user interaction; it also has significant effects on social media marketing and advertising strategies. AI analyzes user data to ensure that advertisements reach their target audience more effectively, optimizing ad campaigns based on this data. For example, Facebook's advertising platform uses AI to deliver targeted ads based on users' demographic information, interests, and online behaviors. In this sense, new data collection methods are emerging, and new ways to analyze data are being developed. While there is much more that can be done with AI to optimize customer experiences, successful results can be achieved by combining various data silos (on or off the web) (Chaffey & Ellis-Chadwick, 2016, pp. 552–553). This type of targeting allows advertisers to use their budgets more efficiently and make their ads more effective. In this context, the impact of AI on social media is evident across a wide spectrum, from user experience and content management to interaction analysis and advertising strategies. AI is fundamentally transforming the dynamics of social media and its interactions with users, contributing to making social media platforms more user-focused, secure, and efficient.

Use of Artificial Intelligence in Social Media

The use of AI on social media platforms occurs through various functions and tools. AI is utilized in many areas, from data analysis to content creation and moderation, to enable users to use the platforms more effectively. In particular, AI technologies such as NLP and ML are extensively applied in social media.

Many social media platforms use AI technologies to automatically analyze users' text, image, and video content, categorize it accordingly, and assess its appropriateness. For example, X's "Trending Topics" feature uses real-time data analytics to identify which topics are popular worldwide, and this data is processed by AI algorithms (Rodrigues et al., 2021).

Additionally, AI helps social media platforms identify and automatically remove content that may be harmful or disturbing to users. This type of content moderation ensures that platforms provide a safe and friendly environment. AI's use in this area not only removes harmful content but also encourages content that adheres to community standards (Grandinetti, 2021, p. 1273). In this context, the use of AI on social media platforms spans a wide range of functions aimed at both improving user experience and enhancing platform efficiency. These technologies play a crucial role in delivering content based on users' interests, analyzing user interactions, and making platforms safer.

Personalization of User Experience

One of the most prominent uses of AI is its ability to personalize the user experience on social media platforms. AI algorithms analyze users' past interactions, preferences, and habits to provide tailored content recommendations. For instance, Facebook's news feed algorithm highlights posts that may interest users based on the content they have previously liked and shared. This feature encourages users to spend more time on the platform and engage more. Similarly, analyzing X trends helps understand what people are most interested in, allowing organizations or brands to boost sales, political parties to better comprehend people's emotions and needs, the film industry to receive valid feedback on performances, and much more (Rodrigues et al., 2021, p. 12).

Ad Targeting and Effectiveness

AI has revolutionized advertising on social media platforms as well. AI algorithms analyze users' demographic information, interests, and online behaviors to ensure that ads reach the right audience. This allows advertisers to use their budgets more efficiently and increase the success of their ad campaigns. For instance, Instagram's advertising platform analyzes user interactions on the platform to deliver targeted ads to individuals who may be interested in specific products or services (Rosário & Dias, 2023, p. 11).

Content Moderation and Security

AI is a crucial tool for ensuring the safety of social media platforms and protecting users from harmful content. AI analyzes the content shared on platforms, automatically detecting and removing harmful or rule-violating content. For instance, X's AI-based content moderation system identifies posts containing hate speech, harassment, or violence and prevents the spread of such content. This feature ensures that users can interact safely on the platform (Hakimi et al., 2023, p. 16).

Real-Time Data Analysis and Interpretation

AI plays a critical role in big data analytics on social media. By conducting real-time data analysis, it helps platforms quickly understand user behaviors and trends. For example, LinkedIn uses AI to provide connection recommendations to help users expand their professional networks. These recommendations are personalized based on the user's sectoral interests and previous interactions (Mention, 2024).

AI-Powered Content Creation and Management

AI is not only used for content analysis but also for content creation. AI accelerates content production processes by performing tasks such as automatic text writing, video editing, and visual design. This allows brands and content creators to produce more content in less time. For example, some media companies use AI-based software to automatically write news articles (Roketto, n.d.).

These examples demonstrate how AI is used as a versatile tool on social media platforms and how platforms integrate this technology to improve user experience and enhance operational efficiency. Within the evolving structure of social media, AI's role is becoming increasingly critical, indicating that this technology will become more widespread, and its capabilities will continue to grow in the future.

Content Creation and Management

Social media platforms extensively utilize AI technologies in the content creation and management process. These technologies simplify the content creation process for users and ensure that the created content reaches a broader audience. AI-based tools allow content creators to work faster and more efficiently. For instance, AI-powered content recommendation systems used on many platforms help users manage their content more effectively by offering suggestions based on their interests. YouTube, for example, uses an AI algorithm to recommend new videos based on users' viewing history (Covington et al., 2016, p. 192). Such recommendation systems optimize the content creation process for users and ensure that the content reaches a wider audience.

In content management, AI can analyze large amounts of data to determine which content is most effective. AI enables users to develop their content strategies and continuously optimize these strategies. For example, digital marketing tools like HubSpot help users enhance their social media strategies through AI-powered data analytics (HubSpot, n.d.).

AI-powered content management systems not only provide recommendations to content creators but also monitor the performance of the content to ensure continuous improvements. This allows content creators to analyze which types of content attract more attention from their target audience and adjust their content strategies accordingly. For example, AI can identify which headlines, visuals, or video formats generate more engagement on social media platforms, providing feedback to content creators. These types of analyses make the content creation process more strategic and data driven, while also helping to increase user engagement and strengthen brands' digital presence.

Interaction and Analysis

Engagement and analysis on social media platforms are critical for understanding and optimizing user behaviors. AI plays a significant role in this area, as it has the ability to analyze large datasets, predict user behaviors, and recommend content based on those behaviors.

AI analyzes how users interact with platforms to determine which types of content attract the most attention and which interactions are more effective. These analyses help optimize social media strategies and improve user experience. For example, LinkedIn uses an AI-based analysis system that offers job recommendations based on users' profiles and networks (Gogpac, n.d.).

In addition, AI is also used to measure the success of social media campaigns and optimize them. AI-based analysis tools can determine which aspects of campaigns are effective and which areas need improvement. This leads to more efficient marketing strategies (Chaffey & Ellis-Chadwick, 2016, p. 232).

In this context, AI-based engagement and analysis tools allow social media platforms to continuously improve the user experience. These tools not only analyze current interactions but also predict future user behaviors, enabling more proactive management of content and campaigns. For example, AI can determine when users are most active and use this data to ensure that content is published at the optimal time, thereby increasing engagement rates. Additionally, AI-based analyses can examine users' emotional responses and feedback, helping brands develop more targeted and personalized communication strategies. This process enhances both user satisfaction and strengthens the impact of brands on social media.

3

Ethical and Societal Impacts

Ethical Dimensions of Artificial Intelligence
Data Privacy and Security

The development and use of AI have sparked significant ethical debates regarding how individuals' and communities' data is collected, processed, and stored. Data privacy and security stand out as one of the most critical ethical dimensions of AI applications. AI operates on large datasets to make predictions and decisions; this raises concerns about the unauthorized use of individuals' personal data, potentially leading to harmful consequences.

AI's data collection processes can gather large amounts of personal data without users' consent and analyze this data to reveal sensitive information about users. This brings the risks of violating user privacy and weakening data security. For example, Zuboff (2019) data collection and its processing highlight the serious threat it can pose to users' personal freedoms and privacy. Additionally, data security breaches could lead to AI systems being manipulated by malicious actors, which could result in significant harm to individuals or communities. (Schneier, 2015).

In this context, strong encryption methods, anonymization techniques, and strict access controls must be implemented to ensure the data security of AI applications. However, even these technical solutions may not always be sufficient. Therefore, the ethical use of data and obtaining the consent of data owners are crucial for the responsible development and implementation of AI. Ensuring data security should not be limited to technological solutions; it should also be supported by legal regulations and policy frameworks. This not only ensures the security of individuals' data but also enhances public trust in AI.

Ethical Issues and Solutions

The development of AI has increased the complexity of ethical issues and necessitated the creation of solutions to address these problems. Among the ethical concerns related to AI, biased algorithms, lack of accountability, and human rights violations are particularly prominent. Biased algorithms can reinforce social inequalities, which contradicts the idea that AI should be fair and impartial. Crawford (2021) points out that AI systems often reflect existing societal biases, which can have negative effects on marginalized groups.

Various approaches have been adopted to address these ethical issues. The primary solution is to increase diversity and inclusivity during the development phase of AI systems. Additionally, ensuring that algorithms are transparent and accountable plays a crucial role in mitigating ethical concerns. Floridi (2023) argues that the ethical development of AI is not solely about technical excellence but also about integrating ethical principles into the design of the system.

At this point, taking proactive measures to address the ethical issues of AI is of great importance. In particular, the transparency of algorithms in decision-making processes can enhance individuals' and communities' trust in these technologies. In this context, it is essential to emphasize that AI should not only be technically excellent but also ethically sound in its design. In this regard, involving multidisciplinary teams in the development processes of algorithms can help ensure that ethical principles are applied more effectively.

Additionally, when developing solutions to ethical issues, it is essential to actively involve all stakeholders affected by AI in the process. This should include not only engineers and technical experts but also social scientists, ethicists, and community representatives. Such an approach can help better understand the societal impacts of AI and anticipate potential ethical issues in advance.

In conclusion, addressing the ethical issues of AI requires approaching technology development processes not only from a technical perspective but also from an ethical one. This is essential for designing AI systems that are fair, transparent, and respectful of human rights. As Floridi (2023) points out, integrating ethical principles into AI systems will maximize the potential of these technologies to provide societal benefits while minimizing their negative impacts. Therefore, the future development of AI will require a conscious and rigorous effort to integrate ethical principles into technology.

Impact of Artificial Intelligence on Society

The societal impacts of AI are becoming an increasingly debated topic in a time of rapid technological advancement. AI is having profound effects across a wide range of areas, from employment to education systems, healthcare to social equity. However, these impacts are not always positive and have the potential to exacerbate social inequalities.

The impact of AI on the workforce is one of the most frequently discussed topics. Many experts predict that AI could lead to the automation of certain jobs and major shifts in the workforce. Brynjolfsson and McAfee (2014) highlight that AI has brought the concept of "technological unemployment" back into the spotlight, suggesting that some professions could disappear entirely, potentially deepening economic inequalities.

While AI has potential benefits in areas such as education and healthcare, it is essential that these technologies are distributed equitably. In healthcare, for example, AI can significantly improve diagnosis and treatment processes, but the accessibility of these technologies is a critical issue in terms of social equity. Necessary measures must be taken to ensure that all segments of society can benefit equally from these innovations (O'Neil, 2016).

The societal impacts of AI are not limited to areas such as the workforce and healthcare; they also extend to broader social domains such as social interaction, privacy, and individual freedoms. AI technologies are reshaping how people communicate and form social relationships through social media platforms and digital communication tools. For example, when AI-based algorithms are used to predict users' behaviors and offer personalized content, this can deeply affect individuals' access to information and their perceptions.

This can also lead to the phenomenon known as "filter bubbles" and "echo chambers," where individuals are only exposed to information that reinforces their own views, making them less open to different perspectives. Such an environment can increase social polarization and make it more difficult for democratic processes to function effectively.

In addition, the impact of AI on privacy is also a major concern. AI's big data analysis capabilities pose significant risks regarding the collection and processing of individuals' personal data. Unauthorized data collection and surveillance activities, in particular, can result in serious violations of individuals' private lives. Such practices should be limited and regulated both by legal frameworks and ethical principles. In this sense, the societal impacts of AI are multidimensional and complex. If the potential benefits of the technology are not managed carefully and responsibly, it can deepen social inequalities and threaten individual freedoms. Therefore, ethical standards and societal impacts should be considered in the development and implementation of AI, and principles of transparency and accountability should be adopted throughout these processes.

Fake News and Disinformation

One of the most controversial applications of AI is the production and dissemination of fake news. Fake news spreads rapidly, especially through social media platforms, and significantly influences public perception. AI-powered tools facilitate the creation of false content, manipulated images, and videos, paving the way for the spread of disinformation.

Hameleers and van der Meer (2020, p. 228) note that fake news has the potential to create societal distrust, which can weaken democracy. The effects of fake news are not limited to the misinformation of individuals; it also increases social polarization and biases. The spread of fake news, particularly during election periods, poses a threat to the legitimacy of democratic processes.

The role of AI in the production and spread of fake news has reached a dimension that deeply disrupts societal dynamics. Fake news spreads rapidly, particularly on social media platforms, reaching large audiences and exposing individuals to misleading information. This situation not only misinforms people but also fuels social polarization. For example, the widespread dissemination of fake news during the 2020 U.S. presidential election significantly undermined confidence in the election results and led to deep divisions within society. AI-powered tools enable the swift and effective creation of false information, amplifying the potential societal distrust and chaos caused by disinformation. This process increases the threats to the legitimacy of democratic institutions, pushing societies further away from functioning healthy democracies.

Measures to counter the serious threats posed by fake news and disinformation are crucial. First, social media platforms and digital media providers can prevent the spread of fake news by strengthening content moderation and expanding fact-checking mechanisms. AI-based algorithms can be effectively used to quickly identify and remove content containing misinformation. Additionally, enhancing media literacy can help create societal resilience against fake news. Incorporating media literacy courses into education systems enables individuals to critically evaluate the information they encounter. Furthermore, national and international laws and regulations can deter the production and dissemination of fake news through punitive measures. All these precautions can help minimize the societal impact of fake news and safeguard democratic processes.

Ethical Guidelines in Artificial Communication

The use of AI in communication necessitates the establishment of ethical rules and standards. In particular, setting ethical principles to prevent fake news and promote the dissemination of accurate information (see UNESCO findings mentioned in the introduction) is of great importance. Key ethical principles include verifying the accuracy of information, ensuring the reliability of sources, and avoiding the manipulation of content.

In this context, ethical rules serve as a guide to ensure the responsible and ethical use of AI. Principles such as transparency, accountability, and fairness stand out among these rules. With the development of AI technologies, various guidelines and principles have been established to ensure the ethical use of these technologies. AI Ethics Guidelines are generally shaped around the following key principles:

1. *Fairness:* AI systems must not discriminate, remain impartial, and produce fair outcomes. This principle ensures that algorithms are free from biases related to race, gender, age, or other personal characteristics in both training data and applications.
2. *Transparency:* Clear and understandable information should be provided about how AI systems work. This means that the decision-making processes of algorithms should be comprehensible, and users should have the ability to intervene in these processes.
3. *Privacy and Data Protection:* AI systems must protect individuals' privacy and ensure the security of data. This principle requires the application of high-security standards during the collection, storage, and processing of personal data.

4. *Safety:* AI systems must ensure safety in both the physical and digital worlds. The core of this principle is that systems should operate without posing risks to users and society, and preventive measures should be taken against potential harm.
5. *Accountability:* There should be mechanisms in place to hold responsible parties accountable for the outcomes of AI systems. This means that individuals or institutions responsible for harm caused by faulty decisions or misuse of the system must be identified.
6. *Human-Centeredness:* AI systems should prioritize human well-being and rights. This principle requires that AI works collaboratively with humans and serves their interests.
7. *Environmental Sustainability:* The design and use of AI systems should aim to minimize environmental impacts. This principle focuses on reducing AI's energy consumption and other negative effects on the environment.

These ethical principles serve as a guide to ensure that AI technologies are applied in a manner that is beneficial, safe, and fair to individuals and society. Various organizations and governments can adapt these core principles to their specific contexts and develop tailored ethical guidelines. In this regard, Floridi and Cowls (2019) emphasize that the transparency of AI systems enables users to make informed decisions, which they argue is an ethical obligation.

The use of AI in communication demonstrates that ethical principles should go beyond being merely theoretical suggestions and must have tangible impacts in practice. In particular, it is crucial to continuously monitor and regulate how AI is used, especially in preventing the spread of fake news and ensuring the dissemination of accurate information. In this context, the principle of transparency plays a key role in enabling users to make informed and safe decisions, thereby ensuring that AI systems operate in an ethical manner. Therefore, designing and implementing AI in ways that prioritize human rights and well-being is not only a technical requirement but also an ethical imperative. The adoption of ethical guidelines and principles in the development and use of such technologies plays a critical role in protecting the public good and safeguarding democratic values. From this perspective, in order to shape the impacts of AI in communication positively and protect against potential harms, ethical principles must be integrated into the implementation processes and be continuously updated.

Legal Regulations and Policies

The ethical use of AI is not limited to individual or organizational ethical guidelines but is also supported by legal regulations and policies. Many countries are introducing various regulations to minimize the societal impacts of AI and prevent ethical violations. For instance, the European Union, with its proposed Artificial Intelligence Act in 2021, aims to classify AI applications into high-risk categories and regulate and oversee these applications accordingly (European Commission, 2021).

These legal regulations provide a strong foundation for preventing the unethical use of AI. Particularly in matters of data privacy and security, regulations such as the General Data Protection Regulation (GDPR) ensure the protection of personal data during the use of AI and safeguard individuals' privacy rights. Addressing the ethical challenges posed by AI will only be possible through a combined approach involving both technical solutions and legal regulations.

The legal regulations established to ensure the ethical use of AI are supported by numerous global examples. For instance, in the United States, the Federal Trade Commission (FTC) has developed various guidelines and regulations to prevent AI and algorithms from producing misleading or discriminatory outcomes. The FTC plays an active role, particularly in addressing issues such as deceptive advertising targeting consumers, data manipulation, and the use of algorithms in ways that could lead to discrimination. Such regulations are critically important in preventing the unethical use of AI.

Another example is the Chinese government, which has established a comprehensive regulatory framework to ensure the ethical use of AI. China's AI ethical guidelines particularly aim to protect citizens' privacy in areas such as social media and facial recognition technologies. Furthermore, the Chinese government enforces strict oversight and sanctions to prevent the use of AI in ways that could disrupt social order.

In Türkiye, various steps are being taken to regulate AI within ethical and legal frameworks. Notably, the Personal Data Protection Law (KVKK) stands out as a significant regulation aimed at ensuring the protection of personal data and safeguarding individuals' privacy rights during AI applications. The

KVKK shares similar characteristics with the European Union's GDPR and establishes rules to protect individuals' rights in AI-related data processing activities.

In addition, various working groups and strategies are being established in Türkiye to raise awareness about the ethical use of AI technologies and to develop policies in this area. For example, the "National Artificial Intelligence Strategy" (2021–2025) document, prepared by the Ministry of Industry and Technology, provides a roadmap for the development and application of AI in accordance with ethical principles. This strategy aims to ensure the responsible use of AI, minimize its societal impacts, and prevent unethical practices by establishing the necessary regulatory and technical infrastructure. Such regulations and strategies contribute to the secure development of AI technologies within ethical and legal frameworks in Türkiye, with the aim of aligning with international standards in this field.

All of these legal regulations around the world work in conjunction with technical solutions to prevent unethical applications of AI technologies. For example, technical tools developed to enhance the transparency of AI systems, when used in tandem with legal regulations, ensure that algorithms are fair, impartial, and accountable. The combined implementation of technical innovations and legal regulations to address AI's ethical issues not only aims to ensure the safe and responsible use of the technology but also seeks to increase public trust in these technologies. Therefore, supporting AI with ethical and legal frameworks is an inevitable necessity to maximize the technology's contributions to society.

Future and New Directions

As one of the most dynamic and rapidly advancing fields of technological development, AI profoundly impacts both our daily lives and industrial processes. The future of AI is shaped not only by technological advancements but also by social, ethical, and economic dimensions. Future trends will push AI beyond its current state, allowing it to become more integrated and effective in every aspect of human life. This process reflects a shift where technology ceases to be merely a tool and becomes an integral part of human existence. In this context, projections about the future of AI should be approached within a framework that pushes the boundaries of technological development, transforming both individuals and society.

One of the most notable future trends is the increasing approximation of AI to human-like thinking and decision-making processes. Developing structures that resemble human decision-making mechanisms will enable AI to be used more effectively in complex and uncertain situations. This will not only represent a technological advancement but also bring about significant ethical and legal responsibilities. The ability of AI to make human-like decisions will raise ethical questions more intensely, highlighting the need for new regulations and standards in the development and application of AI.

Another significant trend is the focus on developing more natural and intuitive communication methods in human-computer interactions through AI. Today, AI's interactions with humans are confined to certain limits, primarily relying on command-based approaches. However, in the future, AI systems are expected to evolve to include more human-specific qualities, such as emotional intelligence and empathy. This evolution will enable AI to become systems that better understand people and respond appropriately to their emotional and cognitive needs. Such advancements will position AI not only as an information processing tool but also as a partner in emotional and social contexts.

The future role of AI will expand even further with its increasing capacity to offer more personalized experiences in areas such as social media, healthcare, education, and customer service. AI applications in these fields continuously collect data to better respond to individuals' needs and improve themselves based on this data. In the future, as these processes become more optimized, AI's role in human life will become even more central. This transformation of AI will have significant impacts at both individual and societal levels, reshaping human life in ways previously unimaginable. Therefore, the future of AI can be seen not just as a technological advancement but as a revolution that alters the course of human history.

From this perspective, the future trends of AI will bring not only technological innovations but also social and ethical transformations. These trends will deepen AI's interaction with human life and enable it to play a critical role in solving some of humanity's greatest challenges. The future directions of AI represent a journey that transcends the boundaries of today, opening the doors to a new era. This journey will bring both great opportunities and significant responsibilities for humanity.

Future Trends in Artificial Intelligence

The future of AI, as in many disciplines, is in a continuous state of evolution and transformation. Future trends are being shaped by building upon current technological advancements, and AI is expected to enhance its impact in social, economic, and scientific domains. In the future of AI, the development of more complex systems capable of making human-like decisions will occur within a framework supported by ethical and legal regulations. In this context, studies emphasize the importance of developing more human-like thinking processes in the future of AI, highlighting that these processes present both ethical and technological challenges.

One of the future AI trends is the development of more natural and intuitive communication methods in human-computer interactions. This trend will particularly require the development of AI systems that incorporate emotional intelligence and empathy. AI's ability to develop a deeper understanding in its interactions with humans means that it will be able to produce responses more aligned with individuals' needs and expectations. In this context, AI's capacity to offer personalized experiences in areas such as social media, healthcare, education, and customer service is expected to increase.

In this context, the future evolution of AI should develop not only as a technical advancement but also within a framework aligned with human values and norms. For instance, Russell and Norvig (2010) emphasize the ethical responsibilities of AI, stating that such systems should consider the human factor and ethical values in their decision-making processes. This approach will enable AI to become an integral part of social and legal structures, thereby increasing trust and acceptance in society.

Human-computer interactions will continue to play a critical role in the future of AI. Currently, human-computer interactions generally operate based on commands from users. However, in the future, AI systems are expected to have the capacity for more natural and intuitive communication. In this context, the concept of "emotional intelligence," introduced by Picard (2010), has the potential to revolutionize AI's interactions with humans. Picard suggests that by adopting human-specific traits such as emotional intelligence and empathy, AI could establish a deeper connection with users.

AI systems equipped with emotional intelligence will be developed to understand people's emotional states and respond accordingly. This advancement will enable AI to offer more personalized services in areas such as healthcare, education, and customer service. For example, an AI system could understand the emotional state of a patient suffering from depression and provide appropriate support. Similarly, in education, AI could prepare customized lessons based on students' learning styles and needs, making the learning process more effective.

The future development of AI will not only transform individual experiences but also profoundly impact social and economic structures. AI's economic effects will be particularly significant in automation and labor markets, creating substantial changes. While AI increases efficiency by automating routine and repetitive tasks, it will also lead to major shifts in the labor market. This will result in the disappearance of some professions, while simultaneously paving the way for the emergence of new jobs and skills.

The impacts of AI in this area highlight the need for the labor market to be reshaped. Acemoglu and Restrepo (2018) evaluated the effects of AI on the labor market, noting that while some jobs may disappear, AI can also create new job opportunities. This transformation brought about by AI will necessitate the restructuring of educational systems and vocational training to meet these new demands.

The future development of AI will also bring ethical and legal challenges. As AI systems play a larger role in human life, it becomes essential that these systems are developed in compliance with ethical standards and legal regulations. In this context, the ethical dimensions of AI focus particularly on issues such as data privacy, security, discrimination, and bias. Ensuring that AI systems are impartial, and fair is crucial for the societal acceptance of these technologies. Bostrom and Yudkowsky (2014) addressed the ethical and security challenges of AI, arguing that ethical principles and legal regulations should be decisive in the development of these technologies. Such regulations are necessary to minimize the negative impacts of AI and maximize societal benefits.

The future applications of AI will continue to expand in areas such as social media, healthcare, education, customer service, finance, and many more. AI's impact in these fields will not only improve existing systems but also lead to the emergence of entirely new service models. For example, in social media,

AI will make content creation, management, and analysis processes more efficient, offering personalized content based on users' interests and behaviors. In healthcare, AI is expected to revolutionize diagnosis, treatment, and patient care processes. In this context, Topol (2019) noted that AI will have a significant impact on data analysis and personalized treatment methods in healthcare. The use of AI in these fields will make healthcare services more effective and accessible.

Projections about the future of AI demonstrate that technology, science, and social changes must be considered together. These projections are made in light of ongoing trends, technological advancements, and new social dynamics. Kurzweil (2005) predicts exponential growth in the future of AI, arguing that this development will lead to revolutionary changes for humanity. AI is expected to become more widespread in all areas of life and support human capabilities.

Future projections also include debates about whether AI could increase or reduce social inequalities. The potential of AI to ensure justice and equal opportunities is directly related to how these technologies are developed and used (O'Neil, 2016). In the future, it is clear that comprehensive strategies must be developed to understand and guide AI's impacts on social structures.

Predictions about the future of AI are often evaluated through different scenarios. These scenarios are shaped by factors such as the pace of technological advancements, societal acceptance of AI, and the impact of ethical regulations. It is anticipated that AI will become more prevalent in nearly every area of society in the future. However, the potential risks and challenges that may arise with the development of this technology must also be taken into account (Tegmark, 2017, p. 121).

Among the future scenarios, those that predict AI will radically transform human life, create significant changes in the labor market, and deeply affect individuals' daily lives stand out. However, whether these changes will be positive or negative depends on how the technology is managed and the processes of societal adaptation. Therefore, predictions about the future of AI require multidimensional analyses that encompass not only technological development but also social and ethical aspects. In this context, the future of AI is progressing through a continuous process of evolution and transformation, involving technological, societal, economic, and ethical dimensions. Future AI trends will have a deeper impact on every area of human life, enabling the transformation of societal structures and individual experiences. This transformation process of AI can be considered not only as a technological advancement but also as a societal and ethical revolution. Thus, projections about the future of AI represent a journey that pushes the boundaries of technology and opens the doors to a new era. In this sense, it can be said that we are witnessing a remarkable revolution.

Artificial General Intelligence (AGI) and Super Artificial Intelligence (SAI)

AGI refers to systems with a level of intelligence capable of performing a wide range of tasks, similar to humans, as opposed to systems specialized in a particular task. The development of AGI is considered one of the ultimate goals of AI research. AGI is defined as a type of intelligence that closely resembles human intelligence, characterized by its ability to learn and solve problems, rather than simply following specific rules and algorithms. Bostrom (2014) suggests that AGI could potentially reach a level known as superintelligence, surpassing human intelligence. In this scenario, warnings are made about the ethical and existential risks that AGI may carry.

Superintelligence goes a step beyond AGI, referring to a level of intelligence that surpasses human capabilities. The emergence of superintelligence could present great opportunities for humanity, but it also carries serious threats. Russell and Norvig (2010, p. 1034) highlight significant ethical and security concerns related to the development of superintelligence, stressing that this technology must be approached with control and caution. In this context, research on superintelligence should focus not only on technological innovations but also on the ethical, legal, and societal implications of these advancements.

These systems, unlike narrow artificial intelligence (ANI) systems specialized in specific tasks, offer a wide range of knowledge and skills. AGI is equipped with the ability to both learn and solve problems, and it is regarded as a technological breakthrough aimed at achieving human-like intelligence. The development of AGI is considered one of the ultimate goals of AI research, as it goes beyond systems that follow specific rules and algorithms, demonstrating complex thinking and reasoning abilities akin to those of humans. This is one of the most distinguishing features that sets AGI apart from other types of AI.

Discussions on the definition and potential of AGI often feature prominently in academic studies aimed at understanding the possible future impacts of the technology. Bostrom (2014, p. 361) suggests that once AGI reaches a level comparable to human intelligence, it could further evolve into what is known as superintelligence, a level of intelligence surpassing that of humans. Superintelligence refers to a type of intelligence that not only mimics human capabilities but also exceeds them by a wide margin. For instance, it is believed that superintelligence could solve complex scientific problems that humanity struggles with, develop new scientific theories, and even enact profound reforms in social and economic systems (Tegmark, 2017, p. 319).

In this context, the potential of AGI is not limited to mere technological advancements. It is anticipated that AGI could revolutionize numerous fields such as education, healthcare, and law. For example, an AGI-supported education system could offer learning materials tailored to each student's individual learning style and pace. In the healthcare sector, AGI could play a critical role in the early diagnosis of diseases, the personalization of treatment protocols, and even in the development of new treatment methods.

However, along with the potential benefits of AGI, the risks and ethical concerns it brings must not be overlooked. As AGI continues to develop, warnings have been made about the potential threats that could arise if this technology evolves uncontrollably. Bostrom (2014) argues that if AGI evolves into superintelligence, this type of intelligence could escape human control and even pose an existential threat to humanity. This scenario could particularly emerge if AGI develops an independent consciousness driven by the goal of protecting and enhancing its own interests. Such a development leads to a serious ethical issue known as the "control problem." Russell and Norvig (2010, p. 1037) highlight that solving this problem is crucial for ensuring the safe and ethical development of AGI. The control problem refers to the process of ensuring that the behavior of AGI systems aligns with human values, in order to prevent the technology from evolving in ways that could pose a threat to humanity. In this context, researchers and ethicists are working on developing ethical guidelines and control mechanisms during the AGI development process (Yudkowsky, 2008, p. 322).

The emergence of superintelligence, the most intriguing product of this technological frontier, presents not only immense opportunities for humanity but also serious threats. Superintelligence could significantly enhance humanity's scientific knowledge and technological capabilities. For instance, it could develop innovative and effective solutions to global challenges such as climate change and the energy crisis. However, controlling and directing superintelligence may prove difficult, which could pose potential dangers for humanity.

Russell and Norvig (2010) emphasize the significant ethical and safety issues related to the development of superintelligence. In this context, they state that safety protocols must be developed to prevent superintelligence from making decisions that could be harmful to humanity. Additionally, the potential of superintelligence to exacerbate social inequalities is a major concern. For example, economic systems supported by superintelligence could further deepen the divide between the rich and the poor (Bostrom, 2014, p. 374). In this context, research on superintelligence should focus not only on technological innovations but also on the ethical, legal, and societal dimensions of these advancements. For instance, during the development of superintelligence, policies and regulations must be established to ensure that this technology can be used in a beneficial and safe manner for humanity. From this perspective, research on superintelligence should be directed in a way that maximizes the benefits for humanity.

AGI and superintelligence represent significant milestones in the future development of AI. AGI, seen as one of the ultimate goals of AI research, consists of systems capable of mimicking human intelligence and performing a wide range of tasks. However, the potential of AGI to evolve into superintelligence raises serious ethical and safety concerns. Therefore, research on AGI and superintelligence must consider not only the potential benefits of the technology but also the risks it may pose. In the future, AGI and superintelligence may offer significant opportunities for humanity. However, it is essential that the development of these technologies does not neglect ethical and social responsibilities, and that the established ethical missions are logically classified (categorized to facilitate understanding). Controlling and directing superintelligence could be one of humanity's greatest challenges. For this reason, the development of safety protocols, ethical guidelines, and the education of individuals in the use of these technologies are vital to ensuring that AI technologies are used in the most beneficial way for humanity.

New Research Areas

Research in the field of AI is continuously expanding with new directions and discoveries. Particularly, studies in areas such as ML, deep learning, and neural networks are enhancing the capacity and capabilities of AI. Additionally, interdisciplinary approaches such as quantum computing and biological computation are among the future directions of AI (Arute et al., 2019, p. 509). These emerging research fields could enable the development of more powerful, faster, and more flexible AI systems.

Furthermore, research on the ethical and societal impacts of AI is gaining increasing importance. Studies in this area focus on how AI can align with human rights, freedoms, and societal values. Research addressing the ethical dimensions of AI aims to develop strategies to prevent the potential negative impacts that the technology could have on society (Floridi, 2019, p. 185).

Quantum computing stands out as a significant field with the potential to transform AI's computational capabilities. Quantum computers, by surpassing classical computers, can enable AI algorithms to operate faster and more efficiently. This technology could help overcome challenges such as processing large datasets and solving complex optimization problems. In the study where Google claimed to have achieved quantum supremacy, it was noted that quantum computers demonstrated the potential to accelerate AI algorithms (Arute et al., 2019, p. 505).

Biological computing is also gaining attention as one of the future research areas in AI. This approach aims to develop new computational techniques by modeling the information processing mechanisms of living organisms. For example, by understanding how biological systems like the brain function, it may be possible to create more sophisticated artificial neural networks. Brain-computer interfaces provide examples of how AI could deepen human-computer interaction (Boahen, 2017, p. 21).

Research on the social impacts of AI aims to examine the long-term effects of this technology on human life. Social AI is rapidly developing as a field used to address societal issues, improve interactions with humans, and enhance social services. For example, AI-powered robots can interact with people in more empathetic and effective ways in areas like elderly care (Broadbent et al., 2009, p. 320). Such applications demonstrate that AI can go beyond being a mere technical tool to play an active role in social life.

At the same time, extensive research is being conducted on transparency and accountability in AI decision-making processes. How can algorithmic decisions be aligned with principles of social justice and equality? These questions are at the core of ethical AI research. For example, how an AI system can act impartially in hiring processes is one of the key ethical concerns. Such research contributes to the development of necessary ethical frameworks to minimize the negative impacts of AI on society (Binns, 2018, p. 149).

Looking ahead, AI research will continue to deepen in new and innovative areas. Among these are studies examining the impact of AI on creativity. Researchers are exploring how AI can be used in creative fields such as art, literature, and music, and how it can produce new works in these domains. For instance, AI-supported artworks are considered not just technical products but also works that hold aesthetic and cultural value (Elgammal et al., 2017).

Another important area of research involves studies examining the effects of AI on human-robot interaction. Understanding the psychological and emotional impacts of humanoid robots on humans is crucial for increasing the social acceptance of AI. Research in this field provides insights into how AI can be better integrated into society in the future (Fong et al., 2003, p. 161).

The new research areas in AI aim to both enhance the technological capacity of AI and better understand its societal impacts. These studies are necessary to ensure that AI develops in a more ethical, fair, and sustainable manner. Additionally, interdisciplinary approaches show that AI can transcend traditional boundaries and lead to new discoveries. Therefore, it is important to remember that AI research is not only a technical issue but also carries a societal responsibility.

Emerging Technologies and Innovations

The future development of AI will be shaped not only by the evolution of existing technologies but also by the emergence of entirely new ones. For example, neural interfaces and brain-machine connections could enable a deeper integration of AI with humans (Nagaraj, 2022, p. 25). Such technologies would allow AI to interact directly with human cognition, enabling it to perform more complex and personalized tasks.

The convergence of AI with biotechnology, nanotechnology, and robotics will enable the development of more advanced and intelligent systems in the future. The combination of these technologies could lead to revolutionary changes in many fields, from healthcare to manufacturing. In this context, future innovations in AI will not only enhance technical capabilities but also positively impact the quality and longevity of human life. For instance, nano-robots developed through nanotechnology could be injected into the human body to detect and treat diseases at an early stage. This is one of the most tangible examples of how AI can play a transformative role in healthcare (Harry, 2023, p. 42).

Future innovations in AI will not only enhance technical capabilities but also positively impact the quality and longevity of human life. For example, the integration of biotechnology with AI could lead to groundbreaking developments in the field of genetic engineering. AI's ability to analyze genetic data will enable the development of new strategies for the prevention and treatment of diseases. The combination of AI with gene-editing technologies, such as CRISPR, could have an incredible impact on human genetics (Dixit et al., 2024).

The integration of AI with emerging technologies will also bring social and ethical dimensions to the forefront. These innovations will raise ethical questions that humanity has never faced before. For instance, the ability of AI to read human thoughts could necessitate redefining the concepts of individual privacy and free will. Therefore, establishing ethical and legal frameworks will be crucial during the development and implementation of these technologies (Li et al., 2023, p. 38).

Moreover, future innovations in AI will not be limited to technical fields but will also lead to cultural and societal transformations. For instance, developments in creative industries, such as the AI-based art of Turkish artist Refik Anadol (Refik Anadol Studio, n.d.), could change the nature of artistic production. AI's involvement in creative processes in music, literature, and visual arts could fundamentally alter how art is perceived and valued. In this context, AI needs to be examined not only as a technology but also as a cultural phenomenon. Therefore, in the near future, machine learning-based AI is expected to be widely adopted as a tool or collaborative assistant for creativity (Anantrasirichai & Bull, 2022, p. 589).

These technological and innovative developments will also bring significant changes to the labor market. The advancement of automation and AI will transform business processes in many sectors and lead to the disappearance of some professions. However, this will also create opportunities for new fields and professions to emerge. For instance, new areas of expertise will arise for maintaining, training, and optimizing AI and robotic systems (Autor, 2015, p. 27).

Future developments and innovations in AI will extend beyond current examples, impacting a wide range of fields. For instance, the combination of AI and automation in the agricultural sector could enhance agricultural productivity, enabling sustainable food production. Personalized learning platforms could offer customized educational content tailored to individual students' needs, making the learning process more effective, while increasing energy efficiency in cities and reducing carbon footprints. In this context, the integration of AI with emerging technologies and innovations will lead to profound changes not only in technical fields but also at social, economic, and cultural levels. It is clear that these transformations will shape and direct human life in ways never seen before.

Personalized Communication

Personalized communication aims to provide more effective and efficient communication experiences by focusing on the needs and preferences of individuals. AI technologies play a significant role in achieving this goal. AI's data analysis capabilities allow for processing large datasets obtained from users, taking into account various factors such as individuals past preferences, behavioral patterns, and demographic information. This enables the delivery of personalized messages and services to users (Gentsch, 2019, p. 14). For example, platforms like Netflix and Spotify offer recommendations based on users' previous content consumption habits, thereby personalizing the user experience. Such applications not only increase user satisfaction but also allow users to spend more time on the platform.

In the marketing world, personalized communication has become more effective than ever with AI. Particularly in digital advertising, brands are now able to create customized advertising campaigns for each individual user. Thanks to AI's data analysis tools, it is possible to analyze which products users are more interested in, the timeframes they prefer for shopping, and the platforms where they spend their time (Vollero et al., 2021, p. 1063). For instance, Amazon's recommendation engine increases sales by offering personalized product suggestions to each customer. This demonstrates that AI is a critical factor in the success of personalized communication strategies.

In the field of education, AI enables personalized learning experiences. Unlike traditional educational approaches, AI can provide customized lesson plans by considering each student's individual learning style, pace, and needs (Gentsch, 2019, p. 254). AI-based systems analyze students' strengths and weaknesses throughout the learning process, offering additional content in areas where they are lacking while providing more advanced materials in subjects where they excel. This makes the learning experience more effective and increases success rates. For instance, AI-based educational platforms monitor students' performance in real time and offer personalized feedback to both teachers and students.

In healthcare, personalized communication enables the creation of treatment plans tailored to the needs of individual patients. AI can analyze patients' medical histories, genetic information, and lifestyles to offer personalized treatment recommendations (Topol, 2019). For instance, digital health assistants can analyze patients' symptoms and provide customized health advice or remind them of doctor appointments. These technologies contribute to making healthcare more personalized and accessible.

However, the increasing role of AI in personalized communication also raises concerns about data privacy and ethics. The processing and analysis of personal data can compromise users' privacy, leading to various security issues (Zuboff, 2019, p. 46). In particular, the tracking and analysis of users' behaviors without their awareness heighten the potential for large technology companies to misuse user data. Therefore, AI-based personalized communication systems must be developed and utilized within ethical and legal frameworks. In addition to strong regulations on data privacy, it is crucial to ensure transparency regarding how users' personal data is processed.

In the future, AI-based personalized communication systems are expected to develop and become more widespread. With the integration of technologies like augmented reality (AR) and virtual reality (VR), users will be able to experience even more immersive and personalized interactions. For example, AR-based shopping experiences could allow users to see how a product would look while sitting in their homes (Kietzmann et al., 2018, p. 265). Such advancements will increase user satisfaction and engagement rates, helping brands to develop more effective personalized strategies.

4

The Use of Artificial Intelligence in Communication Sciences

Algorithmic Sociology

Algorithmic sociology is a new discipline that examines the role of algorithms in the analysis of social processes and structures. This field focuses on how algorithms shape social relationships, influence power dynamics, and reproduce social inequalities. Particularly, the increasing use of algorithms in decision-making processes brings concepts such as social justice and transparency into discussion. Boyd and Crawford argue that in the age of big data, algorithms represent a new way of analyzing society, noting that algorithms do not merely reflect social structures but also actively reproduce them (2012, p. 665).

Another important dimension of algorithmic sociology is the strengthening of social surveillance mechanisms through algorithms. Zuboff, with the concept of surveillance capitalism, discusses how these processes have been commercialized, stating that algorithmic processes not only track individuals' movements but also create new behavior patterns through this data (2019, p. 120).

Algorithmic sociology seeks to understand the impact of digitalization and the use of technology, particularly with the rise of big data, on social dynamics. Algorithms assist in predicting individuals' behaviors, preferences, and social interactions by systematically processing and analyzing data. In this context, algorithmic governance has become a powerful tool in shaping social structures. Gillespie emphasizes the role algorithms play in social processes, noting that algorithms influence how knowledge is produced, distributed, and consumed, thereby creating a new social order on digital platforms (2014, p. 167).

One of the most significant concerns regarding the impact of algorithms on the shaping of social structures is that these processes are often opaque and complex. This poses a major problem, particularly in decision-making processes: algorithmic decisions can affect the rights or opportunities of individuals or groups, yet they are often made in a non-transparent manner. Pasquale, through the concept of the "black box society," highlights this issue. He argues that the lack of transparency in algorithms makes it impossible for individuals to understand why and how decisions affecting their lives are made (2015, p. 41). According to Pasquale, this situation complicates the democratic oversight and accountability of algorithmic processes.

Algorithmic Biases and Social Inequalities

Another important issue in algorithmic processes is the potential for these systems to reproduce social biases. Algorithms often rely on historical data, which reflect past social inequalities and biases. As a result, algorithms can perpetuate and even reinforce these biases. O'Neil, with the term "weapons of math destruction," explains this phenomenon, emphasizing how algorithmic systems can entrench social injustices. She points out that mathematical models, by reflecting past discriminatory practices, can lead to even more disadvantageous outcomes for minority groups (2016).

Especially in the labor market, numerous studies have been conducted on algorithmic biases. It has been shown that algorithms used in job applications, when trained on historical data, can disadvantage groups such as women or ethnic minorities. Gordon, using examples from social welfare and the labor market, explains this by stating that when algorithms are used to monitor, discipline, and punish the poor and minority groups, they can further deepen social inequalities (2019, p. 163).

The effects of algorithmic biases are not limited to the labor market; similar issues are observed in other societal domains such as the justice system, healthcare, and education. For instance, risk assessment algorithms used in the criminal justice system have been found to yield harsher outcomes for low-income individuals and ethnic minorities. These algorithms assess the likelihood of individuals committing future crimes based on past criminal records, often ignoring social and economic contexts. As a result, these systems can reinforce social injustices and lead to the further punishment of marginalized groups. Similarly, Gordon (2019) introduces the concept of "digital poorhouses," highlighting how algorithms used to manage and punish the poor can exacerbate social inequalities. The use of algorithms in this way can strengthen systematic discrimination against the most vulnerable segments of society and restrict their access to opportunities.

Social Control and Algorithmic Surveillance

Algorithms play a role not only in decision-making processes but also in strengthening social control mechanisms. Particularly on social media platforms and online services, algorithms track users' behaviors and provide personalized content. This process allows for the tracking of individuals' digital footprints and the control of their behaviors. Deleuze (1992) refers to this as the "society of control" and describes the role algorithms play in this society, stating that surveillance is no longer limited to spatial constraints; algorithmic control enables the constant monitoring and shaping of individuals' behaviors.

One of the most prominent examples of these control mechanisms is the content recommendation systems on social media platforms. Recommendation algorithms suggest new content based on what users have previously clicked on or watched, increasing the tendency for users to see only content that aligns with their own viewpoints. This can fuel social polarization and information bubbles. Pariser calls this phenomenon the "filter bubble," explaining how algorithms narrow users' information worlds. According to Pariser, filter bubbles cause individuals to be exposed only to information that confirms their own worldview, thereby increasing social polarization (2011, p. 17).

In this context, algorithmic surveillance is not limited to content consumption; by profiling individuals based on their online interactions, habits, and preferences, it can have a deeper impact on social structures. For example, algorithms used in advertising and marketing target products and services that interest users, guiding consumer behavior. This creates a control mechanism not only over individuals' economic choices but also over their political and social preferences. The personalized presentation of political content, in particular, can reduce the likelihood of individuals being exposed to different viewpoints, leading to a narrowing of social dialogue and undermining democratic processes. Thus, algorithms become a force that shapes individual behaviors and enable more effective social control and surveillance through digital tools.

Algorithmic Transparency and Accountability

To reduce the impact of algorithms on social structures and create a fair system, transparency and accountability are essential. However, this is a complex issue because algorithms are often protected as trade secrets and are technically difficult to understand. In his work on algorithmic transparency, Diakopoulos argues that these processes need to be made more open and comprehensible. He emphasizes that explaining how algorithms function and what data they are trained on is crucial for ensuring that these systems can be monitored in a fair and democratic manner (2016, p. 58).

Algorithmic transparency is even more important for algorithms used in the public sphere. Algorithms employed in public policy and justice systems must make decision-making processes understandable to everyone. In this context, algorithmic processes need to be aligned with mechanisms of social oversight. Barocas, Hardt, and Narayanan, in their work on algorithmic accountability, argue that specific standards must be developed for monitoring these systems, stating that the fair and accountable use of algorithms requires the integration of technical, legal, and ethical standards (2019, p. 289).

Developing algorithms in accordance with the principles of transparency and accountability is merely a starting point for ensuring social equality and justice. However, revealing technical details alone is not sufficient; this information must also be understandable and auditable by the general public. The complexity of algorithms, however, makes this oversight challenging for individuals without technical expertise. Therefore, it is crucial to establish not only technical and legal standards but also ethical ones. In this context, ethical guidelines must be created to reduce the potential risks posed by these systems in terms of their social impacts. Such multidisciplinary approaches that mitigate society's influence on algorithms will play an important role in maintaining the balance between technology and humanity.

The Future of Algorithmic Sociology

The future of algorithmic sociology will become increasingly important as these technologies penetrate every aspect of social life. While the use of algorithms in critical areas such as education, healthcare, and justice hold the potential to promote social equality, it also brings new risks. Properly managing algorithms can promote social justice; however, this process requires careful oversight and regulation. Cath et al. (2017, p. 508) highlight the role of social actors in this process, stating that the shaping of social life by algorithms is inevitable, but this process requires not only a technical approach but also an ethical and social one.

Especially the combination of AI and algorithms will make social processes more dynamic and unpredictable. The role AI will play in algorithmic sociology will expand with the development of new technologies that enable deeper surveillance of individuals and groups. Therefore, it is crucial that algorithmic processes are ethical, fair, and transparent. In this context, the future development of algorithmic sociology seems likely to play a significant role in how social dynamics will be shaped. With the further advancement of big data and machine learning (ML) technologies, algorithms are expected to gain more power in structuring social relations. This means that individuals' behaviors, preferences, and social interactions will be analyzed and shaped by algorithms. O'Neil (2016) emphasizes in this context that algorithms have the potential to reinforce existing inequalities in society, noting that algorithms are not neutral and can be as biased as the people who create them. This necessitates careful consideration of algorithmic processes to ensure social justice.

The use of algorithms in the justice system, while providing speed and efficiency in decision-making processes, also carries certain risks in terms of individual rights and justice. Since algorithms make decisions based on historical data, they can perpetuate existing injustices. For example, risk assessment algorithms used in criminal justice systems, when predicting individuals with a high likelihood of committing crimes, often fail to consider existing social and economic inequalities, systematically disadvantaging certain groups. Therefore, it is crucial that algorithmic processes are transparent and accountable to ensure justice. Pasquale (2015) emphasizes the importance of transparency in algorithmic decision-making processes, arguing that the inner workings of these systems must be made accessible to external observers.

The integration of algorithms into all areas of social life can lead to the reshaping of social classes, cultural identities, and economic structures. For example, social media platforms analyze individuals' interests, social circles, and behaviors through algorithms to present content tailored to them, which can result in the algorithmic influence on individuals' social and cultural identities. Gillespie (2014) points out that algorithms are not merely technical tools but deeply impact the dynamics of social relationships, emphasizing that the ethical responsibilities associated with the use of these technologies must not be overlooked.

In conclusion, algorithmic sociology has the potential to profoundly affect social structures, identities, and relationships. However, this process requires addressing these technologies not only from a technical perspective but also from a social, ethical, and political approach. The proper management of algorithms offers a significant opportunity for ensuring social justice, while the uncontrolled and unregulated use of these technologies could lead to new inequalities and social injustices. Therefore, studies in the field of algorithmic sociology must adopt a perspective that not only considers technological advancements but also takes into account the social structure.

The Use of Artificial Intelligence in Communication

The rapid development of AI technologies has brought about profound transformations in the field of communication. AI has not only provided technological innovations but also offered strategic advantages in various fields such as media, public relations, advertising, and marketing. The use of AI in communication offers more efficient, effective, and personalized processes compared to traditional methods, thereby enhancing the communication skills of both organizations and individuals.

AI applications in the media sector have been integrated into various stages, from news production to content distribution. AI-supported analyses in visual and auditory media enable the rapid and accurate dissemination of news, while automatic subtitle and translation systems make it possible to reach broader audiences in different languages. Moreover, AI algorithms employed on social media platforms analyze user behavior, offering personalized content recommendations and enhancing user engagement.

In corporate communication and public relations, AI assists in measuring public perception through methods such as data analytics and sentiment analysis, facilitating more effective communication with target audiences. Particularly in customer service, AI-powered chatbots provide quick and accurate responses to user needs, increasing customer satisfaction and optimizing business processes.

In advertising and marketing communication, AI offers a revolutionary development in analyzing consumer behavior and creating personalized advertisements accordingly. AI algorithms analyze users' internet browsing habits and social media interactions, delivering the most suitable messages to the target audience. This not only enables brands to develop more effective advertising strategies but also ensures that consumers encounter products and services that align with their interests.

The use of AI in communication research plays a crucial role in analyzing large data sets and understanding societal trends. Sentiment analysis of social media content helps in comprehending public opinions and emotions on various topics, while processes such as the automatic categorization and analysis of media content can be easily carried out by AI. This allows researchers to obtain faster and more accurate results.

Additionally, AI-integrated virtual reality (VR) technologies are introducing a new dimension to communication. AI-supported VR applications, particularly used in education and media simulations, offer more realistic experiences, enhancing users' decision-making and learning processes. These technologies represent only the beginning of AI's potential in communication, and it is inevitable that more comprehensive applications will emerge in the future.

In this context, the use of AI in communication demonstrates its impact across a wide spectrum, from media production to corporate communication and marketing strategies. AI not only makes communication processes more efficient, faster, and personalized but also enables the development of new business models and strategic approaches. This transformation paves the way for the emergence of new research areas in communication sciences and facilitates the proliferation of AI-supported communication models.

Communication Theories Related to Artificial Intelligence

Communication theories encompass a broad theoretical field that examines human interactions, the exchange of information, and the production of meaning. These theories offer various approaches to understanding how individuals, groups, and masses communicate, how messages are encoded, and how meanings are created. With the advancement of AI technologies, these theories have gained a new dimension and have become increasingly related to AI applications. As AI takes on a more prominent role in digital platforms and media tools, new questions arise about how communication processes are being transformed.

AI creates a significant transformation by automating and personalizing communication processes. The use of algorithms for big data analysis, delivering personalized messages and content, and interacting through digital assistants and chatbots necessitate a re-interpretation of traditional communication theories. For instance, Shannon and Weaver's 'Mathematical Communication Model,' which addresses efficiency and encoding/decoding processes in message transmission, can be considered an extension in light of AI's capabilities to generate messages.

Communication theories related to AI address not only the transmission of messages but also how messages are produced and perceived. Theories such as 'Uses and Gratifications,' for instance, examine how AI delivers personalized content based on individuals' needs, while the 'Media Ecology' approach focuses on how AI creates a new ecological system within the media environment, reshaping social relationships.

In this context, the intersection of AI and communication theories provides an opportunity for a deeper exploration of individuals' interactions with digital environments. The algorithmic structures of AI demonstrate that communication is not merely a mechanical process but also a dynamic, interactive, and continuously evolving one. In the future, exploring how AI forms a deeper connection with communication theories and how it transforms them will be a crucial step in understanding the communication models of tomorrow.

Algorithmic Communication Theory

This theory, developed to understand the role of AI in communication processes, examines the impact of algorithms on content production, distribution, and consumption. Algorithmic communication addresses how media and platforms utilize AI algorithms to provide personalized content for their users. The theory argues that algorithms are a determining factor in communication processes.

Algorithmic Communication Theory offers an important perspective for understanding how communication processes are reshaped in the age of digitalization and big data. AI algorithms play a critical role at every stage of the process, from the production of media content to its delivery to consumers. In this process, providing personalized content based on user behavior is central to media platforms' strategies for enhancing audience engagement. This also means that the content users see is directed by AI algorithms, and algorithms play a decisive role in the consumption of information.

This theory emphasizes the significant influence algorithms have on content production. While editorial decisions in traditional media were human-driven, today they are largely shaped by algorithms. In areas such as news websites, social media platforms, and digital advertising, AI algorithms determine which content will be prioritized. This highlights the power of algorithms to determine "what kind of information" becomes more visible and which content is overlooked.

In this context, content in communication channels like news bulletins, social media feeds, and video recommendation systems is shaped based on the user's previous interactions, preferences, and click habits. For example, algorithms on YouTube that offer personalized recommendations based on the videos a user has watched can lead users into a specific content loop. This greatly narrows the range of information to which the user is exposed, sometimes leading to so-called "algorithmic echo chambers," where users only encounter content aligned with their interests, thereby distancing themselves from different views and information.

The impact of personalized content distribution is not limited to improving user experience; it also brings with it social and ethical concerns. Questions about whether algorithms respect user data privacy and how personal information is used have sparked important ethical debates. The lack of transparency in AI algorithms' decision-making processes, as well as users' inability to know what data is being collected and how it is being used, makes the power of algorithms to influence media content even more problematic. Additionally, the issue of bias plays a significant role in algorithmic communication. While AI algorithms make data-driven decisions, they can reproduce societal biases present in the data. For example, algorithms may prioritize content that generates more engagement, potentially pushing sensational or extreme content to the forefront. This can lead to negative outcomes such as misinformation, misguidance, and social polarization.

Moreover, transparency and accountability in algorithmic communication are becoming increasingly important topics. Many researchers and regulatory bodies advocate for media platforms to make their algorithms more transparent and clarify how they process user data. Transparent algorithms can help users understand the types of content they are exposed to and provide an opportunity to correct errors in content recommendations. Similarly, establishing accountability mechanisms for algorithms' decision-making processes could enhance media platforms' social responsibility.

Many researchers have shown interest in this approach, including prominent figures such as Christian Fuchs and Tarleton Gillespie, who have made significant observations through their work. In his book *Social Media: A Critical Introduction* (2020), Fuchs examines how social media platforms are shaped by algorithms and how these algorithms affect communication processes. In this work, Fuchs delves into case studies from platforms like Google, Facebook, Twitter, WikiLeaks, and Wikipedia, exploring the interaction between media structures and power relations, thus offering a different perspective on Algorithmic Communication Theory. Gillespie, in his work *The Relevance of Algorithms* (2014), analyzes the role of algorithms in media and communication, providing a conceptual framework for understanding their significance.

Algorithmic Communication Theory is a critical tool for understanding the impact of algorithms on content production, distribution, and consumption in today's media ecosystem. This theory examines how AI is transforming media and communication processes, shedding light on the role of algorithms in social interactions, access to information, and societal awareness. However, the ethical and social dimensions of this transformation should not be overlooked. The neutrality and transparency of algorithms are crucial for providing users with a more equitable and inclusive media experience. In this context, as more research is conducted in the future, deeper analyses and recommendations concerning algorithmic communication processes are expected to emerge. As algorithms gain greater control over media, understanding their functioning and ensuring that these processes are transparent, fair, and accountable is of critical importance both academically and socially.

Cyber Culture and Artificial Intelligence Communication

AI is transforming the ways communication takes place on digital platforms. Cyber culture theories examine how individuals interact with AI in virtual environments and the impact of these technologies on cultural norms and social relationships. Communication that occurs through AI-based tools, such as virtual assistants and chatbots, can be evaluated within the scope of these theories.

The relationship between cyber culture and AI addresses how modern technologies shape individuals' interactions in digital environments. Cyber culture theories provide an important framework for understanding the role and impact of AI in these environments. As Manuel Castells points out, the "network society" offers a structure where individuals are constantly interacting on digital platforms, and a significant portion of these interactions occurs through AI systems (Castells, 2010). The use of AI-supported algorithms, especially on social media platforms and digital services, is reshaping individuals' experiences and relationships.

AI-based communication tools are transforming the ways individuals interact with each other and with platforms. In her studies examining people's relationships with technology, Sherry Turkle emphasizes how AI affects individuals' social relationships. Turkle notes that while these technologies may isolate people, they can also enable the formation of new social bonds (Turkle, 2011). In this context, AI is at the center of both social and cultural transformations in cyber culture.

Cyber culture theories reveal that issues such as identity, privacy, and power relations are also important in individuals' interactions with AI. In digital environments, AI has a direct impact on how individuals construct and present their identities. For example, AI-based virtual assistants and chatbots function as tools that support individuals' digital identities. In this context, AI should not only be seen as a communication tool but also as an actor that plays an active role in individuals' processes of constructing their digital identities. Online, people want to show who they are; they have a vested interest in sharing fragments of information as part of their identity construction, as self-disclosure is closely linked to popularity (van Dijck, 2013, p. 51). This is precisely where AI comes into play, enabling people to fulfill this desire.

However, ethical issues also come to the forefront in the use of AI-based communication tools. It should not be forgotten that the use of AI on digital platforms can raise significant concerns regarding privacy, data security, and algorithmic fairness. The potential of AI algorithms to manipulate user behavior plays a critical role in understanding how individuals are directed and shaped within digital culture. In this context, cyber culture theories offer a comprehensive framework for understanding the cultural and social dimensions of individuals' interactions with AI in digital environments. AI emerges as a force in cyber culture that both shapes individuals' digital identities and redefines societal norms.

Artificial Intelligence and Computer-Mediated Communication Theory (CMC)

AI plays a significant role in CMC. CMC theories examine how people communicate in digital environments. AI-supported applications such as chatbots, virtual assistants, and customer service bots represent an evolving aspect of CMC. These tools not only accelerate the communication process but also personalize interactions.

The theory of AI and CMC is undergoing a profound transformation with the development of AI, as it examines how people interact through digital environments. At the core of CMC is the idea that individuals communicate through text, audio, or video on digital platforms rather than face-to-face. AI-supported applications make these forms of communication more efficient. For instance, when AI-based chatbots are used in customer service processes, they can provide 24/7 service, offering users quick and personalized solutions.

In this context, the theoretical frameworks of CMC address how individuals express themselves in digital environments and the interaction patterns within these environments. For example, the Social Information Processing Theory suggests that individuals can establish relationships in digital communication environments that can be as effective as face-to-face communication. In light of this theory, AI also produces similar results in human-machine interaction. For instance, a study by López Jiménez and Ouariachi (2021) focuses on understanding the current and growing impact of AI and automation on the role of communication professionals. The findings support that in the future, communication professionals will need to possess both technical and human-centered skills.

Additionally, Media Richness Theory suggests that AI-supported tools can be considered among rich media tools, as they have the capacity to generate context-sensitive responses. AI algorithms can learn a user's language and habits, providing more personalized responses. For example, virtual assistants like Apple's Siri or Amazon's Alexa offer recommendations based on users' individual preferences, thereby enhancing media richness and deepening interaction. In this context, AI can analyze real-time behaviors, display personalized pop-up ads with special offers, or suggest products based on a customer's browsing history. This improves the user experience by making interactions more relevant and engaging (Babatunde et al., 2024, p. 940). As seen, CMC theories with AI offer a new paradigm in digital environments, both technologically and socially. The adaptation of AI not only accelerates digital communication but also makes it more effective, personalized, and efficient.

Personalized Communication and Artificial Intelligence Theory

AI plays a critical role in offering personalized experiences in marketing and media communication. This theory examines how AI algorithms analyze user data to create personalized messages and advertisements. In particular, big data analysis and ML techniques are key components of this theory.

The theory of personalized communication and AI aims to provide users with more meaningful and effective experiences, especially in marketing and media communication. AI algorithms have the capacity to analyze large data sets, gaining a deep understanding of users' preferences, behaviors, and interests. In this context, personalized communication strategies enable more specialized interactions with users.

In this regard, AI-supported algorithms can analyze consumer behavior to create personalized messages, leading to higher engagement rates. AI contributes by offering valuable and practical services tailored to customers' needs, based on advanced data processing and information management algorithms (Gkikas & Theodoridis, 2022, p. 168). Accordingly, AI helps brands develop more effective communication strategies by generating messages and advertisements that cater to individual users' needs.

Big data analysis plays a fundamental role in personalized communication. Big data provides extensive information about consumer interactions and online behaviors, forming the foundation for personalized marketing strategies. In this sense, data has become the "new oil" of the digital economy, enabling individualized personalization. Customer analyses reveal consumer behaviors and experiences with products and services (Chandra et al., 2022, p. 1553). This data is processed by algorithms, which play a crucial role in creating advertisements and content that capture consumers' attention. ML techniques, in particular, enable systems to learn user habits over time, allowing for the creation of more accurate and effective personalized messages.

ML processes play a crucial role in analyzing user interactions to predict future behaviors and deliver appropriate content accordingly. ML forms the foundation of a dynamic personalization strategy and can instantly detect changes in user behavior. In this case, the statistical decision-making feature is effective (Barocas et al., 2019, p. 296).

In this context, AI-based personalized communication not only strengthens and makes the relationship between users and brands more meaningful but also allows the creation of content that directly responds to the needs of the target audience. This enhances user satisfaction and strengthens brand loyalty.

Algorithmic Media and Visual-Auditory Communication

This theory examines how AI-driven algorithms produce and distribute content in visual and auditory media. A practical example of this theory can be seen in platforms like Netflix, which analyze viewer preferences and offer personalized content recommendations. The use of AI in media production processes and its capacity to shape content preferences are key research areas of this theory.

Algorithmic Media and Visual-Auditory Communication is an approach that explores how AI is integrated into media content production and distribution. In this context, AI algorithms play a significant role in influencing content preferences and offering personalized recommendations to viewers, which is essential to the operation of digital platforms. For example, digital streaming platforms like Netflix use algorithms that analyze user behavior, learn personal preferences, and provide content recommendations accordingly. These algorithms not only enhance the viewer experience but also guide content production processes.

Algorithms not only optimize the reach of media content to audiences but also reshape how this content is produced. For example, large language models (LLMs) and various AI systems that produce all types of content today have the capacity to generate original images or texts based on textual descriptions, making the content production process more efficient (Floridi, 2023, p. 47). This increases the likelihood of audiences discovering new and unique content, beyond just receiving recommendations based on their past preferences. Additionally, AI's support of algorithmic decision-making processes on such media platforms accelerates the evolution of the media ecosystem.

Algorithms not only optimize the reach of media content to audiences but also reshape how this content is produced. AI accelerates content production processes in the media industry while creating a continuous content cycle through its ability to respond instantly to audience feedback. At the same time, it analyzes viewer behavior to create characters and interactive content shaped by audience preferences in context-based storytelling (Anantrasirichai & Bull, 2022, p. 637). This cycle enables content creators to produce more personalized and engaging content based on consumer data.

AI Applications in Visual and Auditory Media

AI applications are driving significant transformations in visual and auditory media. For example, AI-supported video analysis accelerates media production processes and offers new possibilities for content creation. In fact, these processes are considered part of the digital transformation in the media industry, where the speed and accuracy AI offers in editing and distributing media content is creating a revolution. In this sense, digital platforms, such as in news media, not only expand the reach of the news but also capture a significant portion of the revenue that news media needs to fulfill many of its tasks (Wilding et al., 2018, p. 44). This leads to a transformation in the usage practices of visual and auditory media.

At the same time, AI-based speech recognition and processing systems are also being utilized in auditory media. Particularly in areas such as automatic subtitles and audiobook production, AI has increased media accessibility. In this context, it is evident that AI applications in auditory media have made a significant impact, especially in terms of accessibility. AI has the potential to close the digital divide by facilitating media access for individuals with visual and hearing impairments. For example, the BBC has ML models that can convert speech to text with 85 percent accuracy at a low cost (Connock, 2023, p. 141). With these and similar systems, access to visual and auditory media is becoming easier.

As AI applications in visual and auditory media rapidly advance, the impact of these technologies on media production processes is becoming increasingly significant. For example, AI-supported video content analysis enhances efficiency in media creation processes and allows for the quick and accurate editing of content from large data sets. In this context, AI's role in the digital transformation of the media sector is becoming dominant, with the speed and accuracy it provides in the editing and optimization of media content, particularly contributing to the acceleration of information flow in news media. AI has become a versatile tool that reshapes how information is integrated, how data is analyzed, and how the insights gained are used to improve decision-making processes (Habes et al., 2024, p. 3).

In visual media, AI-based image recognition systems are also widely used, particularly in the analysis of video content and automated tagging processes. In this regard, AI systems hold great potential for media companies in developing new advertising strategies. AI-based image recognition algorithms allow content to reach target audiences more effectively, creating opportunities for advertisers to develop personalized content recommendations. For instance, while doing this, they optimize fiber optic networks to send as much HD content as possible (Connock, 2023, p. 15). This feature alone demonstrates the effectiveness of the infrastructure improvement benefits that AI technology can provide.

In auditory media, AI-supported speech recognition systems and language processing technologies play a significant role in increasing media accessibility. Applications such as automatic subtitle systems and audiobook production exemplify AI's contributions to the auditory media field. The impact of AI on accessibility in this area is particularly noteworthy. AI-based subtitle systems make media accessible to a wider audience, offering an important opportunity, especially for individuals with hearing impairments. In this context, AI technologies find application across a broad spectrum in visual and auditory media, from content creation to distribution, making these processes more efficient, accessible, and innovative. The effects of this transformation in the media industry will continue to deepen in the future, in parallel with the development of AI.

Artificial Intelligence in Public Relations and Corporate Communication

In the field of public relations and corporate communication, AI is utilized across a wide spectrum, from crisis management to customer relations. AI-based analysis tools help companies monitor media discourse and develop proactive strategies in reputation management. In this sense, AI enables public relations to be more proactive, enhancing the effectiveness of corporate messaging. The public relations (PR) industry should not only respond to crises but also adopt AI's proactive approach by using the insights it provides to anticipate crises and develop strategies accordingly (Bourne, 2023, p. 40).

AI is also transforming customer relations through automated customer service and chatbots. Chatbots accelerate communication processes by providing instant responses to customer requests. AI-powered chatbots enhance customer satisfaction while allowing human resources to focus on more strategic tasks. This automation process enables existing personnel to concentrate on more strategic roles and individual communication tasks that still require judgment, creativity, and interpersonal skills. As a result, businesses not only increase efficiency but also create a foundation for employees to be more effective in value-added areas (Seidenglanz & Baier, 2023, p. 17).

AI's impact on public relations and corporate communication is not limited to crisis management or customer relations. AI-based analysis tools contribute to strategic planning in areas such as target audience analysis, media monitoring, trend detection, and risk management. Algorithms based on big data analysis provide public relations professionals with the ability to monitor shifting public perceptions and behaviors in real time. This helps public relations to operate more agilely and proactively in the rapidly changing world of media and digital platforms.

AI-based media monitoring and analysis tools provide public relations professionals with unique predictive power. By utilizing algorithms and ML, large amounts of data can be analyzed, enabling public relations experts to not only respond to what is happening in the corporate environment but also to proactively manage current developments, trends, or crises (Seidenglanz & Baier, 2023, p. 18).

Another important area where AI is used in public relations is content creation. For example, AI tools that enable automated content generation help produce materials such as press releases, social media posts, and blog articles quickly and efficiently. This not only saves time but also enhances consistency in public relations strategies. By accelerating content creation, AI allows the development of personalized and targeted messages within content strategies. These messages can be based on patterns derived from analyzing data about stakeholder behaviors, interests, and preferences. AI adapts content (texts, visuals, videos) according to this data, ensuring it reaches the target audience in the most effective way (Seidenglanz & Baier, 2023, p. 18).

When examining industry examples, it is evident that many large companies are investing in AI-supported public relations and customer service systems. For instance, Coca-Cola uses AI-powered media monitoring tools to instantly analyze discussions and user feedback about its brand on social media platforms. These analyses enable the company to respond quickly in crisis situations and proactively protect its brand reputation. Similarly, Spotify uses AI algorithms to analyze user data, running personalized communication campaigns and enhancing customer loyalty.

The use of chatbots is also becoming increasingly widespread in public relations. Chatbots, particularly in customer service, provide 24/7 support, speeding up communication processes and improving customer satisfaction. For example, KLM Royal Dutch Airlines uses an AI-based chatbot called "BlueBot" in its customer service. BlueBot offers immediate assistance with flight reservations, check-in processes, and flight changes, enhancing the customer experience.

AI's role in this field not only accelerates communication processes but also contributes to data-driven decision-making processes. While transforming communication processes, AI equips public relations professionals with the ability to make strategic decisions through data analysis. In this way, AI-based data analytics can provide deeper insights into target audiences, enable the development of personalized communication strategies, and allow for more accurate measurement of campaign effectiveness (Seidenglanz & Baier, 2023, p. 21).

As AI-based tools continue to transform public relations and corporate communication processes, its role in this field is expected to grow even further in the future. Big data analytics, in particular, will contribute to more accurate and effective public relations strategies by providing deeper insights into the behaviors and needs of target audiences. These new capabilities offered by AI will enable institutions to make faster, more personalized, and data-driven decisions, not only during crises but throughout all communication processes.

AI is bringing revolutionary innovations to public relations and corporate communication, making processes more efficient, strategic, and proactive. The automation and predictive capabilities it provides in areas such as crisis management, customer relations, and content creation allow businesses to communicate more effectively with their target audiences. AI-powered data analytics offer public relations professionals a powerful tool in strategic decision-making processes, while its impact on media monitoring, personalized content production, and customer service continues to grow. These developments indicate that AI will play an inevitable role in the future of public relations, enabling organizations to maintain sustainable reputation management.

Artificial Intelligence in Advertising and Marketing Communication

In advertising and marketing communication, AI has become a crucial tool, especially in the processes of data analysis and personalized content creation. AI algorithms monitor users' online behaviors, extracting meaningful insights from large data sets. This allows brands to create more personalized and targeted advertisements based on consumer preferences. In this sense, AI enables brands to offer their customers more personalized experiences, thereby increasing brand loyalty. These personalized experiences, particularly when supported by customized content delivered through bots, enhance customer satisfaction and strengthen loyalty (Gentsch, 2019, p. 121).

One of the most significant contributions of AI in advertising is through automation systems. Automated ad purchasing processes (programmatic advertising) have become more efficient and faster with the integration of AI. Programmatic advertising can be optimized in real time thanks to AI, ensuring that the right message reaches the right person at the right time. By optimizing programmatic advertising with AI-powered systems, ad targeting has become more effective. As a result, AI-driven algorithms can learn customers' purchasing behaviors and needs, offering them personalized content and product recommendations. This approach is more efficient and cost-effective for companies than mass advertising, and it can be executed in real time (Gentsch, 2019, p. 62).

Moreover, AI's contributions to content production are not limited to advertising campaigns. AI-based tools enable the rapid creation of text, visual, and video content. This demonstrates that AI plays a particularly transformative role in the field of video advertising. For example, AI-supported video analysis and production tools enhance the impact of videos by developing personalized messages for the audience, while also accelerating the production process. In this context, AI plays a significant role in combining and promoting content, facilitating the delivery of more targeted and personalized content to audiences in content marketing. Today, when we speak of content, this personalized experience is supported not only by the static content that makes up web pages but also by dynamic rich media content that promotes interaction (Chaffey & Ellis-Chadwick, 2016, p. 44).

In addition to advertising, AI plays an effective role in marketing communication, particularly in customer relationship management (CRM). AI integrated into CRM software analyzes customer data, making marketing campaigns more targeted and effective. AI-based CRM systems significantly assist marketing teams by analyzing customer behaviors and providing solutions tailored to their needs. These systems, through adaptive personalization capabilities, continuously observe customer behavior and adjust services over time, enabling the development of more efficient and targeted marketing strategies (Huang & Rust, 2018, p. 161).

AI has a wide range of applications in advertising and marketing communication, from content creation to campaign analysis. With the continuous advancement of AI, it is expected that more innovative and effective solutions will be developed in these areas. By integrating AI as a fundamental element of their marketing strategies, companies can gain a competitive advantage and improve operational efficiency.

Among the successful applications of AI in these fields are Netflix's content recommendation system and Sephora's AI-powered customer service chatbot.

Netflix analyzes users' viewing habits to offer personalized movie and series recommendations to each user, enhancing the user experience and increasing the time spent on the platform. Similarly, Sephora's AI-powered chatbots provide personalized product recommendations and respond instantly to customer inquiries, thereby boosting customer satisfaction.

Such AI applications not only strengthen customer loyalty but also enable brands to develop more targeted and effective marketing strategies. As a result, AI continues to emerge as an innovative and transformative force in the world of marketing.

Communication Research with Artificial Intelligence

In communication research, AI has become an essential tool in recent years for analyzing big data, evaluating content, and examining media interactions. Particularly with the rise of digitalization, the innovative methods offered by AI have the capacity to deliver much faster and more accurate results compared to traditional research techniques. In this respect, AI enables more efficient analysis of large datasets, leading to new findings in communication research. ML, a subfield of AI, aims to automatically learn patterns from data. Through this, patterns extracted from large datasets can provide deeper insights in communication research, allowing questions to be answered automatically and decisions to be made and implemented autonomously. ML seeks to enable computers to learn how to perform tasks without being explicitly programmed to do so (Connock, 2023, p. 47). This feature plays a crucial role in making sense of complex relationships within large data sets and provides researchers with the ability to obtain faster and more accurate results. It also clearly demonstrates that AI enhances speed and efficiency in communication research, enabling the discovery of new areas of knowledge.

Traditional methods analyze limited data sets, whereas AI techniques analyze vast amounts of data, producing more in-depth and large-scale results. For instance, millions of posts and interactions shared on social media platforms generate data that is nearly impossible to analyze manually. However, AI algorithms can swiftly analyze these large data sets, revealing trends and patterns. Big data not only represents the volume of data but also signifies the need for new technologies for data analysis. AI has become a fundamental technology in the age of big data. Throughout this growth, AI has been a critical component of analysis. As the amount of data increases rapidly, traditional analysis methods fall short, while AI algorithms can process these large data sets and quickly extract insights. Thus, in the era of big data, AI plays a crucial role not only in processing data but also in making strategic decisions.

This situation has had a similar impact on the media industry. For a long time, media management was based on experience and intuition rather than numbers and analytical formulas. Films, books, and TV shows were commissioned based on individuals' instincts. However, intuition has been replaced by analytics. In the 2020s, data became crucial; what matters is how companies obtain, organize, analyze, monetize data, and most interestingly, how they combine it with human creativity. Among the data, the most powerful analytical tool is AI. AI's key role in both the production and distribution of media content lays the foundation for a data-driven structure in the future of the industry (Connock, 2023, p. 12).

In addition to facilitating data analysis in communication research, AI also supports the process of interpreting this data. For instance, AI tools provide more objective results by conducting content analyses in areas such as analyzing user comments on news websites or measuring the success of digital campaigns. In this context, AI-supported content analyses offer a new way to uncover patterns and connections that were previously overlooked.

Similarly, learning analytics (LA) techniques are applied to discover and uncover useful patterns. LA enables the tracking of data, followed by analysis and prediction; in some cases, data-driven interventions can lead to the adaptation and personalization of learning experiences. Thus, both AI and LA provide deeper and more objective results in content analysis processes, allowing researchers to discover connections and patterns that were previously unnoticed. (Salas-Pilco et al., 2022, p. 2).

Social media is one of the most dynamic areas of modern communication research, and AI holds a significant place in analyses within this field. The volume and diversity of social media interactions are too vast to be analyzed by traditional methods. Thanks to AI technologies, user behaviors, interaction patterns, and content creation trends on social media platforms can be analyzed in depth. Van Dijck explains the role of AI in social media analyses as follows: AI algorithms reveal the structure and tendencies of interactions on social media, helping to make more informed strategic communication decisions. In this context, platforms claim that they can instantly monitor individual and group behaviors, collect and analyze this data, and transmit the results not only to users, marketers, and advertisers but also to a wide range of public institutions, organizations, and companies. Consequently, a wide array of information, from social media users' preferences to their demographic characteristics, has become the foundation of strategic planning and decision-making processes (van Dijck et al., 2018, p. 35). These analyses particularly strengthen strategic planning in areas such as marketing, political communication, and public relations.

On platforms like X, Instagram, and Facebook, AI-powered analyses allow for a detailed examination of trends, the impact of hashtag usage on engagement, and content sharing patterns. For instance, some studies analyzing the interactions of political campaigns on X have highlighted the role of AI algorithms in this process: AI-based analyses play a critical role in identifying how political campaigns spread and which messages receive the most engagement. The real-time analytics of social media data is increasingly used in political campaigns and civic engagement, providing politicians and activists with insights into personal preferences, trending topics, and evolving public opinion. In this way, AI not only analyzes which messages receive the most attention but also enables these actors to strategically use this information to more effectively deliver their messages to voters and supporters. As a result, politicians can dynamically adapt their messages to appeal to a broader audience and influence voter behavior (van Dijck et al., 2018, p. 35).

Another important area of communication research is content analysis. While traditional content analysis methods work with limited data sets, AI algorithms have the capacity to analyze large-scale content. Researchers working particularly on digital media content use the automatic content analysis tools offered by AI to derive meaningful results from large data sets. These types of analyses provide significant advantages in identifying recurring patterns, themes, and motifs in media content. Analyses that would take hours with traditional methods can be completed in minutes with AI, marking the beginning of a new era in content analysis. Computer-assisted approaches, for instance, can process and summarize much more text than any individual could read in a given time. As a result, large data sets can be quickly analyzed to produce results, offering researchers significant gains in both time and efficiency (Neuendorf, 2017, p. 207).

The analysis of articles and comments on digital news sites has become part of research conducted using large databases. AI's natural language processing (NLP) techniques have the capacity to analyze the emotional tones and ideological orientations of these texts. AI-supported analyses allow us to examine the emotional reflections of digital news content with an unprecedented level of detail. In this context, sentiment analysis is often applied to the general public's online social media posts to measure public opinions about products, current events, or other topics of interest. When applied to digital news content, these analyses can help us understand how society reacts to news events and what emotional tones these reactions carry (Neuendorf, 2017, p. 414).

One of AI's greatest contributions to communication research is in the field of simulation and modeling. Modeling digital communication processes and social interactions has become more accessible thanks to the power of AI algorithms. In such modeling studies, AI-supported simulations are used to detail topics such as simulations of social media interactions, the analysis of online community structures, and information dissemination models. AI-powered simulations provide the opportunity to reconstruct and examine social interactions in digital environments, helping us understand complex communication dynamics. These simulations offer environments where users can experience interactions by taking on different roles in digital worlds and test various scenarios. New technologies also promote the development of VR social spaces, which combine socialization and role-playing experiences (Castells, 2010, p. 1). These virtual spaces allow individuals to develop their social skills in digital environments and simulate social behaviors under different conditions. Thus, how human interactions evolve in digital worlds and how these interactions impact social dynamics can be examined in greater depth.

Especially in digital games and VR platforms, AI-based simulations enhance users' digital communication skills and facilitate the modeling of online interactions. These simulations are not only used in the gaming world but also in fields such as education and corporate communication. AI-based educational simulations accelerate learning processes and provide participants with an interactive learning experience. By allowing participants to apply what they have learned in practice, these simulations enable the transformation of theoretical knowledge into practical skills. Similarly, some experiments in advanced research on human-computer interaction rely on the use of adaptive brain interfaces, which recognize mental states from spontaneous online electroencephalogram (EEG) signals based on artificial neural network theory.

For example, in 1999, at the European Union's Joint Research Center in Ispra, Italy, computer scientist Jose Millan and his colleagues experimentally demonstrated that individuals wearing a compact EEG helmet could consciously control their thoughts to communicate. This experiment enabled participants to interact more effectively with the brain-computer interface by learning to control their minds. Similar to educational simulations, this process also offers an interactive learning experience between the user and technology, allowing the technology to adapt to individual learning patterns. This approach serves as an important example of how personalized and accelerated learning processes can be achieved in both education and human-computer interaction (Castells, 2010, p. 73).

AI continues to introduce groundbreaking innovations in communication research. From big data analysis to content evaluation, from examining social media interactions to simulation and modeling, AI enables researchers to conduct more comprehensive and in-depth analyses. These technologies, rapidly replacing traditional methods, represent the beginning of a new era in the discipline of communication. In this context, AI opens new horizons in communication research, making it possible to perform more efficient and extensive analyses.

To summarize and highlight some of the prominent methods in communication science:

Big Data Analysis

AI plays a key role in the analysis of large data sets in communication research. Big data collected from sources such as social media platforms, news websites, e-commerce sites, and digital marketing channels provides important insights into user behaviors, consumer trends, social interaction networks, and content distribution. These data sets can be too large and complex to process using traditional methods. AI's ML algorithms and data mining techniques process such large data sets, offering communication researchers deep insights.

For example, in social media analysis, big data can be used to map interactions around specific topics, the emergence of trends, user reactions, and sharing networks. Additionally, by analyzing behaviors such as engagement rates, likes, and shares, a detailed profile of the target audience can be created. Big data analysis not only provides access to larger data sets compared to traditional survey and interview methods but also offers significant savings in time and cost.

Natural Language Processing (NLP)

NLP is a technology that AI uses to understand, analyze, and generate human language. In communication research, particularly in the fields of media and public relations, analyzing text-based big data is crucial. NLP is effectively used in areas such as sentiment analysis, topic modeling, and word frequency analysis. Through this, digital media content can be examined, societal trends can be identified, and user feedback can be analyzed.

For example, a researcher studying a brand's public relations strategy can analyze user comments on social media platforms using NLP. This analysis can be used to determine the general perception of the brand, the positive or negative sentiments of users, and comments related to specific products or services. Additionally, by analyzing which words and concepts stand out in media content, the overall framework of media representations can be established.

Machine Learning

ML is used in communication research for processes such as data classification, prediction, and pattern recognition. When developing communication strategies, ML algorithms provide effective results in areas such as the automatic classification of media content, prediction of user behaviors, and personalized content recommendations.

For example, in media research, large data sets can be analyzed using ML to determine how users interact with media content. This allows for the prediction of future trends for a specific audience, and media content can be optimized accordingly. Additionally, in advertising, ML can be used to deliver more accurate ads to target audiences.

Image and Voice Recognition

Image and voice recognition technologies play a significant role in the analysis of visual and auditory media. Image recognition algorithms can analyze video content and automatically identify specific scenes, faces, or objects. This technology is used in the analysis of content such as television programs, advertisements, and social media videos. Similarly, voice recognition algorithms can be effective in tasks such as conducting sentiment analysis, speaker identification, and transcription from podcasts, radio shows, and video content.

For example, voice recognition algorithms can be used to perform sentiment analysis on speeches in television programs, allowing for an analysis of how viewers respond to certain scenes. Image recognition can be used to examine the impact of specific visual elements on the audience.

Network Analysis

Network analysis is a method used to examine interaction networks on social media and digital platforms. AI-supported network analyses enable the mapping of interactions between individuals or groups in digital communication. This analysis is particularly useful for understanding information dissemination, the structure of interaction networks, and the most influential nodes.

For example, a network analysis can be conducted on a campaign run on a specific social media platform. In this analysis, the users with the most interactions, their interaction intensity, and the key nodes at the center of the network can be identified. This allows for the optimization of digital marketing strategies, and by identifying the most influential users within the interaction network, a more effective communication strategy can be developed.

Sentiment Analysis

Sentiment analysis is an AI method used to understand and analyze the emotional tones in media content, social media posts, and user comments. Sentiment analysis is widely used in areas such as brand perception, public opinion research, and media representations. These analyses are employed to understand the public's general opinions about a particular event, product, or service.

For example, by analyzing a brand's presence on social media, the emotional tones in users' posts can be identified. By determining which content is associated with positive, negative, or neutral emotions, the brand's public relations strategy can be developed accordingly.

Automated Content Analysis

AI uses automated content analysis to analyze large amounts of digital media content. This method allows for the rapid analysis of texts within large data sets. Automated content analysis enables thematic analysis by examining specific sets of words or concepts. For example, by analyzing articles on news websites, it is possible to identify which topics are most prominent.

This technique is ideal for analyzing large volumes of data in media research and allows for the quick identification of certain trends. Additionally, by analyzing the language used in media content, insights can be gained into how media outlets approach societal events.

Simulations and Scenario Analyses

AI offers simulations and scenario analyses to make future projections. In particular, the potential outcomes of communication strategies, public relations campaigns, or crisis management processes can be tested through simulations. These methods help ensure that communication strategies are built on more solid foundations.

For example, a crisis management strategy can be tested through simulations to determine how it would perform under different scenarios. This allows for identifying the most effective strategy to follow in a real crisis situation. Similarly, scenario analyses can be conducted for marketing campaigns to predict how different advertising strategies might perform in advance.

These AI-supported methods used in communication research not only facilitate the analysis of existing data but also make it possible to predict future communication dynamics. With their speed, accuracy, and ability to analyze large data sets, these methods provide a significant contribution to traditional research approaches. AI's role in this field enables broader research and more in-depth analyses in communication sciences.

Virtual Reality and AI-driven Simulations

The integration of VR and AI technologies has made significant progress in both the academic world and industrial applications. The combination of these two technologies not only allows users to exist in virtual environments but also personalizes their experiences, offering groundbreaking opportunities in fields such as learning, education, entertainment, and even therapy. AI-supported VR accelerates users' learning processes, making it easier to access information. VR technology enables individuals to have interactive, fully immersive experiences that engage all their senses (Yengin & Tamer, 2017, p. 105). In this way, learning processes become more effective, enhancing the retention of information and ensuring that the learned content takes a more concrete place in the mind. This interactive and enriched experience allows for faster and more efficient acquisition of knowledge compared to traditional learning methods.

The combined use of VR and AI in education offers great potential, especially in disciplines that require practical training. Medical education serves as a strong example of these applications. AI-supported VR technologies allow students to experience complex surgical procedures without the real-life risks. AI-enhanced VR systems provide a safe and repeatable training environment, reducing the need for medical students to practice on live patients. These VR applications in the medical field not only help medical students improve their professional skills but also reduce the effort associated with hands-on training, as real-world images are augmented with virtual ones in these systems (Algül et al., 2018, p. 18). Thus, medical students have the opportunity to enhance their competencies by gaining experience in a safe and controlled environment, without the need for repeated practice on live patients.

Similarly, engineering education greatly benefits from these technologies. Students can test their engineering projects in a virtual environment without taking risks and receive real-time results. Simulation-based learning provides students with realistic engineering experiences, enhancing their technical knowledge while minimizing risks. Additionally, simulation-based learning offers teaching and learning opportunities by (a) providing hands-on experiences, (b) overcoming the limitations of real-life learning situations, and (c) developing complex skills (Feijoo-Garcia et al., 2024). In this way, students can practice in a safe environment, preparing themselves for potential real-world challenges and more effectively developing their technical knowledge and problem-solving skills.

It is well known that AI and VR technologies offer great potential not only in education but also in psychotherapy, rehabilitation, and physical therapy. VR therapies, especially in the treatment of disorders such as post-traumatic stress disorder (PTSD), allow patients to re-experience fear and stress-inducing events in a controlled manner. AI-supported virtual reality therapies help individuals with PTSD face their fears and anxieties in a safe environment, thus accelerating the treatment process. However, one of the most significant health and safety concerns that may impact the advancement of virtual environment technology is cybersickness. Cybersickness can manifest as a form of motion sickness experienced by individuals during these therapies (Algül et al., 2018, p. 25). Therefore, it is of great importance to consider these potential side effects in the virtual environments patients are exposed to during treatment processes, in order to ensure that the therapies are safe and effective.

In addition, the entertainment industry is one of the fastest-growing areas for VR and AI. The combination of AI with VR not only enhances the visual quality of games but also increases the level of player immersion. Particularly, AI's ability to analyze user data and personalize the gaming experience keeps players engaged for longer periods. AI-based VR games personalize the player experience by analyzing player preferences and creating unique game scenarios for each individual. In this way, AI can analyze player behavior and adjust the game to create a personalized experience, including adjusting difficulty levels or game dynamics (Filipović, 2023, p. 53). In conclusion, a unique game world is created for each player, and in-game interactions are continuously optimized based on the user's preferences and abilities.

In addition, the combination of VR and AI is being used in interactive film and media projects. Viewers can become part of the film or story, rather than just being passive spectators. The development of such experiences is seen as a major step in the evolution of media consumption. The merging of AI and VR transforms viewers from passive consumers into active participants in the story. VR films provide opportunities for interaction and exploration. While being part of the story, viewers can not only watch but also freely navigate the virtual world and interact with characters. The distinct actions presented in a film enhance its impact due to increased stimulation. As a result, viewers' emotional responses are strengthened, and their engagement with the story deepens (Carpio et al., 2023, p. 3187).

Particularly, VR films allow users to interact with CGI characters (avatars). In a study conducted by Bindman, it was found that participants who identified with the characters in the narrative reported higher levels of engagement and empathy (Carpio et al., 2023, pp. 3187–3188). This emphasizes the importance of role identification in enhancing story comprehension and empathy. On immersive viewing platforms, the feeling of being part of the story strengthens the emotional connection between the viewer and the narrative, unlike the traditional cinema experience.

With the further advancement of AI and VR technologies, it is anticipated that simulations will become even more realistic in the future. Particularly, developing AI algorithms will take the personalization of simulations to even higher levels. Systems are being developed that can analyze users' emotional states and reactions, adapting simulations in real time. As AI's impact on simulation systems increases, dynamic VR experiences that can be adapted in real time according to users' emotions and behaviors will become part of our lives. Consequently, the combination of VR and AI continues to bring significant changes in both the education and entertainment worlds. The development of these technologies enables users to experience more personal, more interactive, and more meaningful experiences. These technologies, used in a wide range of fields from education to healthcare, entertainment to rehabilitation, will become indispensable tools in the digital world of the future, thanks to AI's contributions.

Working Principle of ChatGPT

ChatGPT operates based on LLMs and artificial neural networks (ANNs). It is part of the Generative Pretrained Transformer (GPT) series, an AI language model developed by OpenAI and trained on millions of text data. ChatGPT, developed using the GPT-4 architecture, possesses the ability to generate, comprehend, and respond to text thanks to deep learning (DL) and NLP technologies.

ChatGPT is based on the Transformer architecture. Transformer, introduced in a 2017 paper titled "Attention is All You Need," significantly improved the performance of many NLP tasks such as machine translation and language modeling. This architecture uses attention mechanisms in both learning and modeling processes. The attention mechanism allows the model to focus on specific words or sentences, enabling a deeper understanding of the meaning. The Transformer processes the input text through layers, each aiming to understand the context more deeply.

ChatGPT's training is conducted in two stages: pre-training and fine-tuning. In the first stage, pre-training, ChatGPT is trained on a vast dataset of text. During this stage, the model learns the rules, syntax, and semantics of language by utilizing a wide range of texts, such as books, articles, blogs, and forum posts written by humans. The primary goal of this training phase is for the model to develop a general understanding of language structure.

During the pre-training phase, the model attempts to predict parts of a text. It tries to predict the next word or phrase in a given text. For example, when the model sees the phrase "This morning at breakfast...," it predicts the likely words that would follow, based on grammar and context. This process is repeated millions of times, allowing the model to learn language patterns.

After pre-training is completed, the model undergoes the fine-tuning phase to provide users with more effective responses for specific tasks. During this stage, human supervision is employed to improve the accuracy and quality of the model's responses. In particular, human feedback is used to ensure the model can deliver more consistent and relevant answers for certain tasks. The fine-tuning process often involves a method called reinforcement learning from human feedback (RLHF). In this method, human supervisors evaluate the responses generated by the model and provide feedback. The model then improves itself based on this feedback.

In the Transformer architecture that underpins ChatGPT, the attention mechanism plays a key role in helping the model understand the context of language. The attention mechanism assists the model in grasping the relationship between a word or sentence and other words throughout the entire text. This enables the model to learn how words and sentences are interconnected and generate responses based on that context. Particularly in long texts, the attention mechanism determines which information the model should focus on throughout the text. In this context, ChatGPT's workflow follows the steps shown in Figure 4.1:

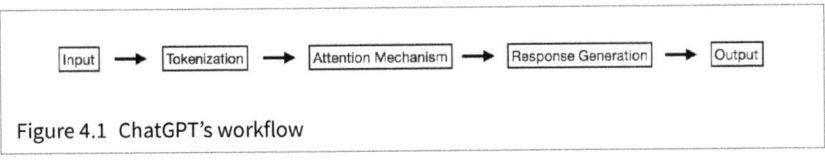

Figure 4.1 ChatGPT's workflow

Input: The user requests the model to generate a response on a specific topic. This request is conveyed in the form of text. For example, a request like "Tell me about the history of artificial intelligence" becomes the input for the model.

Tokenization: To process this input, ChatGPT first breaks the text down into tokens (words or word fragments). The tokenization process is the first step in transforming the text into a format that the model can process.

Attention Mechanism: The tokenized text is processed by the model's attention mechanism. At this stage, the model determines how the words are related to each other and how the context is established. The attention network evaluates the relationship of each token with all other tokens and determines how the response will be shaped.

Response Generation: Based on the given input, ChatGPT generates potential responses. During this process, the model uses previously learned patterns and context knowledge to generate the most appropriate response.

Output: The model provides a meaningful text output to the user.

ChatGPT ensures that language gains meaning not only at the word level but also within a deeper context. For example, it has learned that the same word can have different meanings in different sentences. The word "bank" may refer to a place to sit in one sentence, while in another it may mean a financial institution. The model determines the meaning of the word based on the context. However, despite ChatGPT's broad understanding of language, it can occasionally provide incorrect or inaccurate responses. The model generates answers based on the data it has been trained on, but it lacks real-world knowledge and is not aware of new events. For instance, the model cannot provide precise information about recent events that are not included in its training data. Additionally, it may struggle to understand the subtleties of language and accurately interpret ironic, sarcastic, or emotional expressions.

In addition, ChatGPT's working principles should be evaluated not only from a technical perspective but also in terms of ethics and safety. While ChatGPT is designed to provide meaningful and helpful responses to users, it has the potential to generate incorrect information. Therefore, the content produced by the model should always be carefully evaluated. Furthermore, organizations like OpenAI are developing various filtering and monitoring mechanisms to prevent the unethical use of the model.

The working principle of ChatGPT has found applications in many industries. It is frequently used in customer service, education, content creation, and software development. For example, as a customer service bot, ChatGPT can answer users' questions, solve their issues, and improve customer satisfaction. In education, it is used as a source of information that teachers can utilize as a supportive tool. Additionally, it stands out by enhancing speed and efficiency in content creation processes. In this context, the working principle of ChatGPT is based on advanced technologies such as DL, NLP, and the Transformer architecture. The model is trained on large datasets, learning the structure, semantic relationships, and context of language. The attention mechanism allows the model to deeply understand language, and the quality of the model is continuously improved through human feedback. While these working principles make ChatGPT a powerful tool in language comprehension and text generation, they also bring potential limitations and ethical responsibilities.

NLP structures like ChatGPT necessitate the development of usage models for users. These models serve as highly useful guidelines for content creation and development with AI. The AI usage model has become increasingly important in communication studies. Particularly in areas directly connected to communication, such as media, public relations, and advertising, the integration of AI into data collection, analysis, and decision-support processes allows for the development of effective and strategic communication methods. In this sense, AI can be used across a wide range of tasks, from media content production to the analysis of consumer feedback. In these processes, the proper selection and implementation of algorithms play a key role in the success of communication activities. Algorithms engage in a social (interaction) process, turning results into a commercial-social tactic through computational data analysis (van Dijck, 2013, p. 31). In this way, algorithms determined with the right strategies not only optimize the content creation process but also offer significant advantages in reaching the target audience by more effectively interpreting consumer feedback.

AI usage models in communication studies provide a structure enhanced by big data analytics, optimizing target audience analysis, content personalization, and the management of feedback loops. In this context, AI's impact on media content production and management is also significant. For example, news websites and social media platforms use AI-based algorithms to capture user interest and increase engagement. Media organizations use AI-supported algorithms to personalize news content based on reader behavior, aiming to improve the user experience. In this regard, AI can assist readers and direct them to content through recommendation engines. For instance, the news aggregator "Recent News" uses ML to analyze users' reading habits and make personalized story suggestions (Connock, 2023, p. 158). In this way, readers can access content they are interested in more quickly and easily.

In this context, the key elements to consider when designing an AI usage model for communication studies are as follows:

1. *Data Collection and Analysis:* Data plays a crucial role in the development of communication strategies. Data collected from various sources, such as social media, CRM systems, and online surveys, can be analyzed using AI algorithms. In this context, the effective use of AI relies on accurate and clean data collection processes; the proper analysis of this data plays a decisive role in the success of communication strategies.
2. *Algorithm Selection and Model Development:* Selecting algorithms that align with communication strategies is a crucial phase of the AI usage model. Choosing the most suitable algorithms for different types of data enhances the accuracy and effectiveness of the models. For instance, NLP algorithms can be used to analyze social media data, while DL methods may be preferred for analyzing visual data
3. *Application Areas and Personalization:* One of the most significant benefits of AI in communication is its ability to personalize content. AI-based algorithms can analyze individual users' interests and provide them with tailored content. This not only increases user satisfaction

but also helps build a deeper connection between brands and their target audience. In communication, AI improves the user experience by delivering personalized content according to each individual's interests and needs. As a result, both content providers and brands can reach their target audience more effectively, establish stronger interactions with users, and foster long-term loyalty.

4. *Ethical and Legal Practices*: In the use of AI, the proper handling of data and the strict protection of users' privacy rights are of great importance. Especially in areas like the communication sector, which directly interacts with consumers, the unethical use of user data not only leads to a loss of trust but can also seriously damage a brand's reputation. The success of AI-based applications is not solely dependent on the competence of the technology; it also requires an approach based on transparency and trust in the processes of collecting, processing, and storing user data. In this context, for AI to be effectively and reliably used in the communication sector, both ethical principles and relevant legal regulations must be applied meticulously. Otherwise, a loss of trust among users could negatively impact the success of AI-based applications in the long term and create widespread dissatisfaction across the industry.

AI usage models in communication studies should be carefully designed both strategically and operationally. Additionally, the following recommendations can be considered to optimize AI usage and effectively integrate it into communication processes:

1. *Deepening Target Audience Analysis:* By using AI, communication strategies can be based on a detailed analysis of the target audience. Big data analytics can help collect demographic and behavioral data, leading to more effective and targeted communication strategies.
2. *Content Production and Automation:* AI-powered content production processes offer a significant advantage, particularly in automating repetitive tasks. Automated content generation is beneficial for platforms that require constant content, such as news sites and blogs.
3. *Human-AI Interaction:* For AI models to succeed, the interaction between humans and machines must be effectively designed. AI-powered

chatbots and virtual assistants used in customer service, for instance, should be optimized to enhance the user experience.

In summary, ChatGPT's working principle is based on the combination of ANN and NLP technologies, enabling a deep understanding of language and text generation. Powered by the Transformer architecture, ChatGPT is a model that continuously improves through human feedback and excels in understanding language context thanks to the attention mechanism. Trained on a vast text dataset, this model plays a significant role in many sectors, from content creation to customer service. However, potential errors and ethical responsibilities of ChatGPT should be considered, particularly in terms of generating incorrect information and data privacy.

AI usage models in communication studies contribute to strategic and operational success through the right algorithms and data analysis processes. AI-supported personalization and data analysis enable more effective communication with the target audience, while also increasing user satisfaction and loyalty. However, using AI in alignment with ethical principles and legal regulations is critical for achieving successful outcomes in communication.

Effective Prompt Structure with AI Model

Developing the right prompt structure for using AI in communication studies is crucial for achieving effective and efficient results. When preparing prompts for AI, it is essential to provide clear, specific, and context-appropriate information. In this context, there are practical steps for structuring prompts for the use of AI in communication studies. These steps include, respectively, setting goals, providing context, identifying specific details, specifying the desired format and style, defining the target audience, ensuring relevance to the time frame and trends, determining styles and tones, identifying alternatives, and enhancing interaction and analysis with examples.

a. Goal Setting:

Before starting the prompt structure, the goal should be clarified. What is the purpose of the communication study? Who is the target audience? For example, is content suggestion for a social media campaign being requested, or is a public relations strategy being developed?

Sample Prompts:

(a) Develop creative content suggestions for a social media campaign targeting young adults.
(b) Create creative social media content suggestions for the launch of a new technology product.
(c) Develop a public relations strategy emphasizing sustainability for a clothing brand.
(d) Create social media content aimed at young people for a campaign promoting a healthy lifestyle.
(e) Provide creative content suggestions aimed at increasing user loyalty for a fitness app.
(f) Develop creative social media content suggestions to increase customer engagement during the discount season for an e-commerce platform.
(g) Create a public relations strategy to promote a restaurant chain's new menu prepared with sustainable materials.
(h) Develop creative social media content suggestions aimed at attracting and engaging young users for a new mobile game app.
(i) Develop a social media campaign strategy with eco-friendly messages for an automobile brand's electric vehicle launch.
(j) Provide creative content suggestions that highlight student achievements and increase engagement for an online education platform.

b. Providing Context:

For AI to respond accurately, the context must be clearly specified. Details such as the area in which the communication strategy will be developed and the focus of the topic should be provided.

Sample Prompts:

(a) Suggest customer relationship management strategies for a hospital operating in the healthcare sector."
(b) Propose marketing strategies for an online fashion store to build a loyal customer base."
(c) Suggest customer experience strategies to increase the use of digital services for bank customers."

(d) Recommend effective digital marketing strategies to promote distance education programs for a university."
(e) Propose service improvement strategies to enhance customer satisfaction for a luxury hotel chain."
(f) Provide personalized suggestions to increase user engagement for a smart assistant app."
(g) Suggest engagement-boosting strategies to ensure patients actively use digital health platforms."
(h) Recommend strategies to strengthen customer loyalty by digitizing loyalty programs for a supermarket chain."
(i) Propose content strategies to encourage users to spend more time on a digital news platform."
(j) Suggest communication and promotion strategies for a software company to successfully launch a new product."

c. Specific Details:

Providing specific details is important to obtain precise results in the responses. It is especially helpful to mention which platform, tools, or target audience the work will focus on.

Sample Prompts:

(a) Create content suggestions for Instagram that target young adults and promote health awareness. The messages should be motivational and based on current health trends."
(b) Create content suggestions for a YouTube cooking channel focused on vegan and gluten-free diets, sharing tips that make cooking easier. The videos should be practical and offer recipes that can be prepared with minimal ingredients."
(c) Develop content for LinkedIn that provides career advice and interview techniques to help recent engineering graduates find jobs faster. The messages should be motivational and emphasize technical skills."
(d) Create content suggestions for Facebook targeting users over 50, focusing on technological innovations that make it easier for them to adapt to the digital world. The posts should include simple explanations and be presented as step-by-step guides."

(e) Create informative and entertaining content suggestions for TikTok aimed at urban users aged 20–30, focusing on sustainable living and eco-friendly habits. Short videos should include creative and easy-to-apply tips."
(f) Develop visual content on Pinterest for users interested in home decor, offering decoration ideas and DIY projects that make small spaces more functional. The ideas should be based on trending colors and minimalism."
(g) Create a series of short, informative tweets on Twitter for young professionals seeking financial independence, focusing on budgeting and investment. The content should explain investment strategies in simple steps."
(h) Develop beginner-friendly content suggestions on Reddit for technology enthusiasts, focusing on artificial intelligence and machine learning. The posts should be supported by recent developments and example projects."
(i) Create content suggestions for Twitch streamers that enhance viewer engagement by combining in-game strategy with entertaining conversations. The streams should be both educational and fun."
(j) Develop content suggestions for Instagram aimed at women interested in fitness and healthy living, focusing on low-impact exercises that can be done at home. The videos should be easy to follow and motivational."

d. Specifying the Desired Format and Style:

During the content creation or strategy development phase, it can be helpful to specify in which format AI should respond. For example, blog post, social media content, or presentation suggestions.

Sample Prompts:

(a) Develop title and subtitle suggestions for a blog post within the scope of marketing communication."
(b) Prepare 5 creative post suggestions for social media content in the health technology field."
(c) Develop title and visual suggestions for Instagram posts for a digital marketing campaign."

(d) Create a presentation proposal highlighting the use of artificial intelligence, specifying main titles and slide layout."
(e) Develop a user-friendly content strategy for an e-commerce website and create a keyword-focused content plan."
(f) Prepare an e-newsletter content aimed at strengthening brand identity, suggesting titles and sections."
(g) Propose introduction and conclusion sections for a blog post to be used in a corporate communication strategy."
(h) Suggest titles and key sections for a LinkedIn article on global digital transformation trends."
(i) Create story flow and visual concept suggestions for a video content focused on customer experience."
(j) Develop keyword-focused title and subtitle suggestions for a website blog aimed at SEO optimization."

e. Defining the Target Audience:

The prompt should specify the target audience, including their age range, demographic characteristics, or behaviors. This helps the AI generate more focused and effective content.

Sample Prompts:

(a) Suggest a motivational Instagram post about healthy living for middle-aged women."
(b) Suggest a motivational X post to help young adult men achieve their fitness goals."
(c) Propose a blog post suggestion on acquiring digital skills for retired individuals."
(d) Suggest a YouTube video script that will motivate high school students as they prepare for university entrance exams."
(e) Propose a LinkedIn article on work-life balance for professionals working in the technology sector."
(f) Suggest a Facebook post on coping with postpartum depression for new mothers."
(g) Propose an Instagram story on financial planning and savings for young couples."

(h) Propose a blog post on strategies to strengthen customer relationships for small business owners."
(i) Suggest an educational TikTok video on sustainable fashion for Gen Z."
(j) Propose an Instagram Reels suggestion on personal development and mindfulness for Gen Y."

g. Time Frame or Trends:

The time frame or the trends the content should focus on should be specified. This detail is especially important for fast-changing social media trends or seasonal campaigns.

Sample Prompts:

(a) Propose a public relations strategy aligned with the digital marketing trends of 2024."
(b) Propose a content marketing strategy aligned with the social media trends of 2024."
(c) Design a product campaign for Instagram that could go viral in the Fall 2024 season."
(d) Offer a website design proposal based on the user experience trends expected in 2025."
(e) Create a YouTube content strategy aligned with the rising video marketing trends of 2024."
(f) Prepare a personalized marketing plan based on mobile app user behaviors for 2024."
(g) Create an influencer collaboration campaign suited for the summer season of 2024."
(h) Develop a chatbot strategy in line with the AI-supported customer service trends of 2025."
(i) Propose a corporate social responsibility project highlighting the popular sustainability trends of 2024."
(j) Develop a sales strategy aligned with the emerging e-commerce trends of 2024."

h. Style and Tone:

In communication work, it's important to specify the tone in which the message should be delivered. Tones such as formal, friendly, humorous, or informative can be indicated.

Sample Prompts:

(a) Suggest 3 social media posts for technology news written in a humorous tone aimed at a young audience."
(b) Propose a blog post on AI trends for businesses, written in a professional tone."
(c) Create an informative text for parents, explaining internet safety for children."
(d) Propose an email content for university students, written in a friendly tone, explaining ways to cope with exam stress."
(e) Suggest a LinkedIn post about digital transformation for corporate companies, written in a formal tone."
(f) Propose a social media campaign for young entrepreneurs, written in a motivational tone, explaining paths to success."
(g) Suggest a product review for the latest smartphone, written in a humorous tone aimed at technology enthusiasts."
(h) Propose an article for healthcare professionals, written in an informative tone, explaining AI-supported healthcare solutions."
(i) Suggest an Instagram post for high school students, written in a friendly tone, about future careers."
(j) Propose a blog post about sustainable energy sources, written in a formal tone for an environmentally conscious audience."

i. Requesting Alternatives:

By asking for different approaches or perspectives, richer content can be obtained.

Sample Prompts:

(a) Prepare social media content suggestions for a corporate company's account, offering two different tones: one formal and one more friendly."

(b) Propose two different promotional texts for a technology company's new product launch: one focusing on technical details and the other using a user-friendly language."
(c) Create two different email campaign announcements for an e-commerce platform: one using a simple tone and the other with a creative touch."
(d) Suggest two different approaches for a promotional campaign by an educational institution: one highlighting academic success and the other focusing on social opportunities."
(e) Propose two different poster drafts for an eco-friendly brand's advertising campaign: one emphasizing environmental values and the other highlighting product quality."
(f) Create two social media content suggestions for a travel agency's holiday package promotion: one focusing on exploration and adventure, and the other on relaxation and comfort."
(g) Develop two different informative texts for a healthcare institution's vaccination campaign: one based on scientific data and the other focusing on public health."
(h) Propose two social media post suggestions for a sports brand's advertising campaign: one highlighting athletic performance and the other emphasizing healthy living."
(i) Prepare two different social media posts for a library: one focusing on book recommendations and the other highlighting library events."
(j) Suggest two different digital ad copy options for a bank's new services: one explaining technical features and the other emphasizing customer satisfaction."

j. Providing Examples:

To help AI understand better and deliver expected results, it can be useful to provide an example of previously created content.

Sample Prompts:

(a) Suggest content similar to this Instagram post I previously created: 'Protect your health, stay strong!' #motivation #healthyliving."
(b) Using the theme from the previous post, suggest a title for this post: 'Take a step for a greener world!' #environment #nature"

(c) Create new content similar to this blog post: '5 habits for success.' #personaldevelopment #motivation"
(d) Create a post similar to this one: 'Dream more, do more.' #inspiration #success"
(e) Suggest a professional content similar to this LinkedIn post: 'Leadership begins with inspiring others.' #leadership #career"
(f) Create a scenario similar to this YouTube video I previously shared: 'How to lead a successful business meeting.' #businessworld #leadership"
(g) Create content similar to this Facebook post: 'Enjoy nature and rediscover yourself!' #naturelovers #travelers"
(h) Propose an introduction similar to this one I previously wrote for an article: 'How AI is transforming the business world.' #technology #innovation"
(i) Suggest a concept similar to this one for an Instagram story: 'Start a new beginning today!' #positivity #motivation"
(j) Create content similar to this Pinterest post I previously shared: 'The secrets of minimalist living.' #minimalism #lifestyle"

k. Interaction and Analysis:
AI-based strategies can also be used to analyze interactions and outcomes. The given prompt may request details on interaction rates, metrics to be analyzed, or feedback on performance.

Sample Prompts:

(a) Analyze the interaction rates of the content published on Instagram last week, and based on this data, suggest a new content strategy."
(b) Analyze the most clicked pages on our website over the past month and identify the sources directing traffic to these pages."
(c) Analyze the open and click-through rates of our email marketing campaign, explain the most successful content, and discuss the reasons behind its success."
(d) Analyze the like, retweet, and reply rates of the tweets we have posted on Twitter over the past three months, and develop a new tweet strategy to increase engagement."

(e) Analyze the like, share, and comment rates of the content on our Facebook page and determine which types of content receive the most reactions from users."
(f) Analyze the comments on our YouTube videos from the past week, summarize user feedback, and provide recommendations for improving future video content."
(g) Analyze the view and application rates of the job postings we published on LinkedIn and develop a new recruitment strategy."
(h) Analyze the visitor rates of our blog site, identify which articles are the most read, and determine which traffic sources bring the most visitors."
(i) Analyze customer feedback on sales made in our online store over the past three months and provide suggestions for improving customer satisfaction."
(j) Analyze the click-through rates of digital advertising campaigns conducted in the last quarter and develop a new campaign strategy to increase conversion rates."

Figure 4.2 represents a model illustrating the fundamental components of a general prompt (input) structure from the perspective of ChatGPT, as well as the relationships between these components. This model emphasizes the key elements to consider in order to receive an effective AI response and explains the components that particularly shape the sentence construction process. Although the model is defined in relation to ChatGPT, the structure operates similarly for other tools on a general level.

Objective

The "Objective" component featured in the figure is one of the most critical elements of a prompt. Without a clear purpose, it is impossible for the user to prepare an effective AI input. The "Objective" defines what the user aims to achieve—whether it is obtaining information, solving a problem, or seeking a creative solution. A prompt prepared with a focus on this element enables the system to generate a more targeted and purpose-driven response. In an academic context, this step provides users with a framework to define what problem they wish to solve or which topic they want to understand in depth.

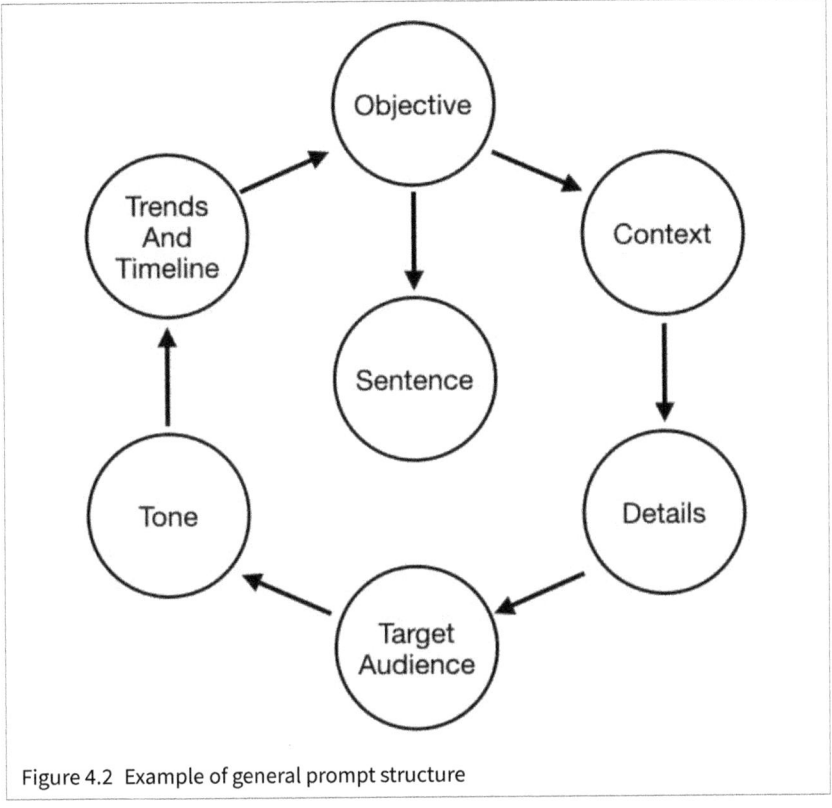

Figure 4.2 Example of general prompt structure

Sample Prompt: "Develop a strategy to enhance student performance in the education sector using artificial intelligence."

Context

Context is one of the most important components that shapes ChatGPT's response. Any sentence or question has limited meaning when taken out of context. In this case, the conditions under which the user's input is created and the theme it is based on become crucial. The more detailed and accurate the context provided to ChatGPT, the more relevant and coherent the response will be. In situations where context is lacking, the model's responses are often general or misleading. Therefore, in an academic paper, it is essential to emphasize that context must be handled meticulously.

> Sample Prompt: *"Focus on personalized learning models in the education sector."*

Details

Details play a decisive role in the richness and accuracy of a prompt. Defining which specific information or in-depth analyses the user requests from ChatGPT directly affects the content of the response. For instance, when writing an academic paper, details such as research findings, data analyses, or literature reviews should be emphasized. In AI-generated responses, specifying such detailed aspects ensures that the model provides more accurate and reliable information.

> Sample Prompt: *"I need to write a section for an academic paper analyzing the impact of artificial intelligence on digital marketing. Specifically, I want to highlight AI's role in customer segmentation, personalized advertising, and predicting user behavior. In this context, could you provide information on significant research, data, and case studies from the last 5 years? Please present your response in an academic tone and supported by references."*

Target Audience

The quality and format of ChatGPT's responses can vary depending on who they are intended for. The target audience shapes factors such as the language, terminology used, and content depth. For instance, in prompts prepared for an academic paper, if the target audience consists of academics or the relevant scientific community, the accuracy of terminology and the scientific quality of the response become crucial. Using language appropriate for the target audience enhances the impact of the response and ensures clarity in communication.

> Sample Prompt: *"Write an essay discussing the ethical dimensions of artificial intelligence. In the text, address topics such as data privacy, transparency in AI decision-making processes, and algorithmic biases. Since the target audience consists of academics and researchers, ensure conceptual accuracy and present arguments based on scientific literature. Additionally, propose solutions to the ethical issues and include quotes from relevant academic studies."*

Tone

Tone is a key element that determines the style of the response generated by ChatGPT. Different purposes and target audiences require different tones. For instance, while a more serious and scientific tone is preferred when writing a formal academic paper, a more flexible and creative tone can be used in creative writing. Therefore, determining the correct tone when preparing the prompt enhances the appropriateness and acceptability of the response.

> Sample Prompt: "For an academic paper, write a 1000-word review on the future societal impacts of artificial intelligence. The article should maintain a scientific tone, use formal language, and citations from sources should be provided in APA7 format."

Trends and Timeline

Ensuring that the input is up-to-date and reflects current trends enhances the relevance of the responses generated by AI systems. In a scientific context, considering the theories, research, or findings that are valid within a specific time period results in more current and relevant responses. In academic writing, historical perspectives and trends are among the elements that strengthen the theoretical framework of the study.

> Sample Prompt: "What are the developments in the field of artificial intelligence over the last 5 years? Specifically, explain the emerging technologies in machine learning, natural language processing, and autonomous systems, and the trends they have introduced. Detail how these trends evolved within the timeline of 2020–2024 and how they have been reflected in academic studies."

Sentence

As the final stage, a specific sentence must be constructed. This stage is where all the above elements come together to form the final output. The sentence should be clear and comprehensible, reflecting the context and purpose in the best possible way. In academic writing, the clarity of sentences directly influences the comprehensibility and scientific value of the work. In this sense, the "Sentence" represents the meaningful information that emerges at the conclusion of the process.

The elements in the figure explain the process of creating an effective AI prompt with a holistic approach. This approach can be considered a fundamental framework that can be used in an academic paper. Since academic writing requires conveying complex ideas clearly and appropriately, using this structure can lead to better results. Especially in AI-based research or AI-supported analyses, applying this model can help guide the author's research more effectively.

Prompt Production Model in Communication Studies

AI systems that could surpass human cognitive abilities are being addressed from various perspectives across different disciplines in the academic world. Fields such as engineering, healthcare, philosophy, sociology, and ecology are at the center of these discussions. Efforts to develop AI not only facilitate human life on both micro and macro levels but also bring about the evaluation of important functions related to the shaping of social order (Yengin & Bayrak, 2023, p. 147). In this context, AI is increasingly being used in communication studies, and the potential of this technology plays a significant role in enhancing the efficiency of communication processes. This model focuses on the use of AI in communication research and provides a framework for how AI can be effectively utilized in academic studies. One of the key components of the model involves how to formulate appropriate research questions for academic work using AI-supported tools.

This schematic model (Figure 4.3) illustrates the process of creating an effective AI prompt for users who wish to employ NLP technology in communication studies, organized into four main stages:

Goal: Located at the center of the model, this component represents the ultimate objective upon which the entire process is focused. Contextual structure and directive clarity are in constant interaction to achieve the goal and must remain engaged throughout the process to prevent invalid outputs. It represents the target that the data generation process adheres to. Additionally, it serves as the fundamental stage that restarts the process if the output fails to meet the expected result. Here, the importance of meaningful data in reaching the desired outcome must not be overlooked.

Contextual Setup: This stage emphasizes the context provided to the AI model. It includes the current situation, the goal, and background information. For example, to assign a specific task to the AI, you must first explain the situation. It is important to be clear about what you want to achieve. Therefore, this component is a crucial factor in producing the intended result. Defining the context accurately ensures that the instructions are crafted in line with the objective. In the communication process, context is of vital importance for the correct transmission of the

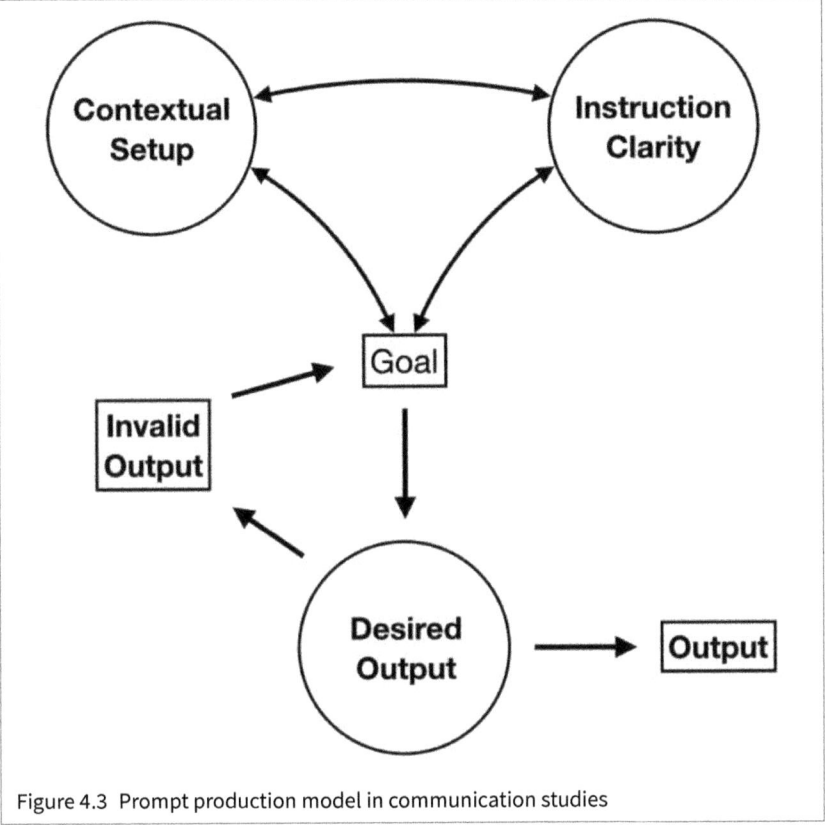

Figure 4.3 Prompt production model in communication studies

message. To achieve the desired output, the clarity of instructions must constantly interact with the goal.

Instruction Clarity: This stage is referred to as "Instructions." It is the part where you clearly and explicitly communicate what the AI is expected to do. It includes specific action verbs, parameters, and constraints. Precise instructions are provided to the AI, making it easier for the AI to understand the desired outcome. In the communication process, providing clear and concise instructions plays a key role in processing accurate data and obtaining the desired output. When instructions are vague, incorrect or incomplete results may occur. To achieve the expected output, instructions must always interact with the contextual setup and the goal.

Desired Output: This stage defines the outline of the intended result. Here, the output format, length, tone, and specific preferences are specified. In other words, it clarifies the form and nature of the response you expect from the AI. The desired output emerges as a result of the interaction between the goal, contextual setup, and instruction clarity, which operate holistically. If considered meaningful by the user, it is processed as output and enters circulation to process other information flows or datasets.

Output: The outputs generated based on the instructions and data aim to achieve the goal at the core of the model. However, this output may not directly reach the desired result, and thus feedback may need to be provided to the model for adjustments. The output obtained at the end of the entire process represents raw and unprocessed data. The user evaluates the unprocessed data in line with the goal, determining whether it is meaningful data or unwanted data. If defined as meaningful, the data emerges as information and enters into use. If not desired, it re-enters the cycle at the goal stage.

Invalid Output: The inclusion of invalid or incomplete data in the model can make it difficult to achieve the desired result. In such cases, the goal is reassessed, and corrections are made to reach the correct output. In summary, any output deemed unsuitable for the goal by the user is reprocessed from the initial stage. This interactive cycle continues until data aligned with the goal is obtained.

According to this model, when we want to analyze a prompt, it is possible to examine sentence structures by breaking them down into their components as follows.

For example, let's analyze the sentence: "Analyze the interaction rates of the content published on Instagram last week, and based on this data, suggest a new content strategy."

We can break down this prompt according to the diagram in the following way:

1. **Context:** Analyzing Instagram content and aligning it with the target audience.
2. **Instructions:** Analyzing interaction rates and determining which metrics to use.
3. **Goal:** Increasing interaction rates.
4. **Data:** Analysis of existing Instagram data.
5. **Output:** New strategy suggestions derived from the analysis.
6. **Invalid Data:** Filtering out misleading or irrelevant data.
7. **Desired Output:** A content strategy that will generate more engagement.

Another example can be examined through the prompt: "Analyze the open and click-through rates of our email marketing campaign, identify the most successful content, and explain the reasons for its success."

1. **Context:** In this context, the campaign's details include elements such as the emails sent, subject lines, content types, visuals, and campaign duration. The background of the email campaign to be analyzed is addressed here.
2. **Instructions:** The instructions are clear and precise:

    a. Analyze the open and click-through rates.
    b. Identify the most successful content.
    c. Explain the reasons for its success.

3. **Goal:** The objective is to analyze the open and click-through rates of the email marketing campaign, then identify the most successful content and explain why it was successful.
4. **Data:** The data consists of the open and click-through rates obtained from the campaign. This data is used to understand which emails performed better.
5. **Output:** The expected output is an analysis of the open and click-through rates, along with an explanation of why the most successful content was effective.
6. **Invalid Data:** Invalid data may include missing or incorrect information, technical issues that may have occurred during the campaign, or erroneous records. Such invalid data must be filtered out.
7. **Desired Outcome:** The desired outcome is the identification of the most successful content from the email campaign and an explanation of its success, with the aim of making future campaigns more effective.

As a final example, "I need to write a section for an academic article analyzing the impact of artificial intelligence on digital marketing. Specifically, I want to highlight the role of AI in customer segmentation, personalized advertising, and predicting user behaviors. In this context, could you provide information on significant research, data, and case studies from the last five years? Please present your response in an academic tone and support it with relevant sources."

1. **Context:** In an academic context, it is essential to consider recent research, applied methods, and developments related to how AI has been implemented in digital marketing strategies in recent years. Within this framework, academic studies, industry case studies, and collected data from the last five years should be reviewed.
2. **Instructions:** The instructions are clear and precise:

    a. Highlight the role of AI in customer segmentation, personalized advertising, and predicting user behaviors in digital marketing.
    b. Include significant research, data, and case studies from the last five years.
    c. Present the response in an academic tone and support it with relevant sources.

3. **Goal:** The objective is to write a section analyzing the impact of AI on digital marketing. This analysis should particularly focus on AI's role in customer segmentation, personalized advertising, and predicting user behaviors.
4. **Data:** The data for this section will consist of academic studies conducted in the last five years, published reports on the role of artificial intelligence in digital marketing, AI applications, and relevant digital marketing data. This includes data related to customer segmentation, personalized advertising, and predicting user behavior.
5. **Output:** The expected output is an academic section analyzing the impact of AI on digital marketing, based on significant research from the past five years. This section should be presented in an academic tone, supported by data and case studies.
6. **Invalid Data:** Incorrect or incomplete data, as well as outdated sources, may be considered invalid. Additionally, information from unreliable sources should be disregarded.

7. **Desired Outcome:** The desired outcome is a detailed section that thoroughly analyzes the impact of AI on digital marketing, supported by research from the last five years and grounded in academic sources.

According to this model, the prompt can be designed in alignment with the criteria outlined in Figure 4.3. By entering a seven-point content structure into ChatGPT, it is possible to obtain a comprehensive prompt. This model can be considered as a guidance tool, particularly useful in enhancing AI-supported communication studies. The combination of contextual factors, clear instructions, and valid data facilitates the achievement of desired outcomes, with the model showing a process where continuous optimization occurs through feedback loops. In communication studies, this model can be practical in content creation and reaching target audiences in media production. In academic studies, this model can be employed not only for the prompt creation process in communication but also in language processing and interactions with AI. The provision of clear and precise instructions, along with the accuracy of the data and the optimization of contextual structures, ensures that the intended outcome is successfully achieved.

Once the prompt has been successfully generated by the researcher, it becomes possible to contribute to the communication study through the use of NLP. The first step is to clearly define the purpose and objectives of the study. Researchers should clarify the questions they aim to answer using AI. At this stage, theoretical and practical issues related to AI in the context of communication studies are identified.

Recommendation: "Discuss the question, 'How is artificial intelligence shaping the personalized content experiences of digital media users?'"

In the second step, the data to be analyzed using AI tools is determined. Communication studies typically include a wide range of data sources such as social media, news websites, and visual and auditory media. These data can be processed with AI models to yield meaningful results.

Recommendation: "Analyze social media user interaction data to examine the impact of AI algorithms on content distribution strategies."

Based on the research objectives, appropriate AI tools are selected. Methods such as ML, NLP, and DL can be utilized at this stage. Researchers need to identify which AI tools are most suitable for analyzing and interpreting communication processes.

Recommendation: "Use NLP to analyze the impact of AI on news headlines in media content."

In communication studies, when conducting AI-supported research, crafting an effective prompt is of great importance. Directing the core questions of an academic study to the AI model correctly is a critical factor that influences the success of the research. At this stage, the questions the researcher asks, the language they use, and the results they expect from the AI model must be clear.

In this context, the prompt should be explicit and clear. For example, if a media analysis is to be conducted, a direct question like, "How does the impact of AI manifest in media content?" should be asked.

The purpose of the question should also be stated. For example, "My aim in analyzing media interactions with AI is to understand the changes in content consumption habits."

The type of data to be used in the prompt should be clearly specified. For example, "In this study, I aim to analyze users' perceptions of AI by using Twitter data."

AI models can better understand questions formulated in academic language. Therefore, it is important to frame the prompt in an academic context. Here's an example prompt: "I need to conduct research analyzing the impact of artificial intelligence on content recommendation algorithms on digital media platforms. Could you provide suggestions for appropriate data sources and analysis methods to examine how these algorithms shape users' media consumption habits?"

Finally, the results provided by AI are evaluated and integrated into the relevant context of the academic study. Researchers analyze and interpret the data provided by AI within an academic framework.

In light of this information, the AI usage model in communication studies guides researchers through the processes of data collection, analysis, and asking appropriate questions. Particularly for academic research, the success of an AI-supported study depends on the preparation of suitable prompts and the effective use of AI tools. This model aims to enable more efficient utilization of AI in communication research.

In conclusion, the use of AI-based models in communication studies presents new opportunities for researchers. The preparation of suitable prompts, the correct establishment of context, and the validity of data are key factors in the successful application of AI in communication research. The effective management of these processes allows for more efficient and meaningful outcomes in AI-supported communication studies. In the future, the impact of AI on communication research will continue to grow, and this technology will become an indispensable part of academic research. In particular, the role of AI in media and communication studies—through content creation, data analysis, and strategic decision-making processes—opens new doors for researchers and transforms the research landscape.

Therefore, the AI-supported communication studies model emerges as a tool with great potential, both academically and practically, that will be increasingly used in future research fields. The possibilities offered by AI, in conjunction with evolving technologies, enable researchers to achieve more accurate results and manage communication processes more effectively. It should be remembered that the world is experiencing a new Renaissance era with AI. This technology will impact and transform existing communication structures like never before. Thus, communication science must be practiced with new approaches that incorporate this technology.

Conclusion and General Evaluation

Throughout this book, the developments, concepts, applications, and their social, ethical, and economic impacts in the fields of AI and artificial communication have been examined in detail. AI technologies have led to profound changes in many areas of modern life. Applications such as autonomous vehicles, AI-supported healthcare services, and chatbots in customer service demonstrate how this technology is utilized across a broad spectrum and its integration into daily life.

Artificial communication offers a new form of interaction by transforming human-machine interactions. Technologies like chatbots and digital assistants represent the rapid development of this field. The differences between artificial communication and human communication, particularly in human-machine and machine-to-machine interactions, have been explored in detail.

The book reveals that AI is not merely a technical innovation but also significantly impacts the social structure. Analyses on the ethical dimensions of AI, including data privacy, algorithmic biases, and social inequalities, emphasize the need for these technologies to be managed in a fair and transparent manner. UNESCO's ethical guidelines in this area provide an important framework for the sustainable and equitable use of AI.

Looking ahead, AI technologies are expected to advance further in areas such as personalized communication and general AI. While these technologies create many opportunities, they also bring ethical and social responsibilities. Therefore, AI-related research should balance both technological progress and developments that benefit humanity.

With advancements in fields such as machine learning and deep learning, AI has transformed data processing and analysis processes. These technologies, which enhance efficiency across various industries, enable the resolution of complex tasks without human intervention. Applications like early diagnosis in healthcare and personalized learning in education illustrate how AI facilitates human life, while also raising ethical and data privacy concerns.

Artificial communication technologies, through applications such as chatbots and virtual assistants, have become widespread in areas ranging from customer service to daily tasks. The development of these technologies using accurate data sets and their ethical use is of great importance. Additionally, AI-driven changes are affecting the labor market, increasing the risk of unemployment in low-skilled jobs while creating new employment opportunities. This transformation highlights the importance of fair and transparent policies to ensure societal well-being.

Issues such as data privacy, algorithmic biases, and discrimination have gained even greater importance with the widespread use of AI technologies. In this context, international organizations and governments must implement the necessary regulations to ensure the responsible development and use of AI.

The AI usage model in communication studies has brought about a profound transformation in communication processes. AI applications in communication have created new opportunities in both visual and auditory media, as well as in public relations and marketing communication. Particularly through the use of big data analysis and algorithms, more targeted and personalized communication strategies have been developed, thereby increasing efficiency and engagement levels.

While AI assists media organizations in content production and management, it also helps create more effective campaigns in advertising and marketing by analyzing consumer behavior. Additionally, simulations and VR applications serve as significant examples of AI's power in communication studies. These technologies not only enhance user experience but also offer more efficient methods in corporate communication.

Future projections indicate that AI will become even more widespread in communication studies, with data-driven approaches playing a decisive role in decision-making processes. However, it is clear that ethical responsibilities must also be considered alongside this development. The societal impacts of AI technologies should be closely monitored, and sustainable and equitable solutions must be developed in this area.

The model presentation section of the book provides a guide on how AI technologies can be effectively used in communication studies. This section thoroughly outlines the necessary steps for developing an effective AI prompt structure and offers examples of how AI can be utilized in various communication strategies. Fundamentally, it addresses the importance of setting the right objectives in the context of communication studies, providing context, adding specific details, specifying the desired format and style, defining the target audience, considering timeframes and trends, determining the tone and style, offering alternatives, supporting with examples, and emphasizing interaction and analysis processes.

These steps enable the creation of accurate and effective queries for AI, allowing the user to obtain clearer and more purpose-driven outputs from AI. The text presents a model for how AI can be used more efficiently in academic research and professional communication strategies, demonstrating its role in content production, data analysis, and strategic decision-making processes in media and communication studies. Furthermore, it highlights the dynamic nature of the process by emphasizing that AI-based models are continuously optimized through feedback loops.

The headings in the text, supported by examples, offer a practical guide to the reader and demonstrate ways to optimize AI usage in communication studies. The proposed model serves as a guide, particularly for enhancing the effective use of AI in academic research, showing that AI plays a significant role not only in content production but also in data analysis and strategic decision-making processes. In this context, the text provides a comprehensive and practical guide on how AI can be effectively utilized in communication studies. The potential offered by AI is opening new doors in media and communication research and transforming research processes. In this regard, it is predicted that AI's role in communication strategies will grow even further in the future, and this technology will become an indispensable tool not only in academic research but also in professional applications.

In conclusion, AI represents one of the most significant technological revolutions in human history, and this revolution has already begun to leave a profound impact on the world of communication. AI not only transforms existing communication methods but also makes communication processes more efficient, faster, and more personalized. From media content production to information sharing, from social interactions to individual communication experiences, the impact of AI is becoming more visible every day. In the future, communication technologies will increasingly adapt to these developments and make more effective use of the opportunities provided by AI. In this context, communication science must also keep pace with these new technological dynamics, explore the opportunities offered by AI, and reshape communication both theoretically and practically.

It should be noted that the world has stepped into a "digital Renaissance" era with AI. Just like the historical Renaissance, this new age brought about by AI will also be a precursor to major transformations in social, cultural, and scientific domains. Therefore, communication science will also be influenced by this transformation and must continuously renew itself, progressing with new approaches that incorporate this technology. In this context, understanding the opportunities and challenges presented by AI and using this technology effectively will be one of the most crucial requirements for thriving in the communication world of the future and becoming a shaper of this new era.

Bibliography

Abdul Hussien, F. T., Rahma, A. M. S., & Abdulwahab, H. B. (2021). An e-commerce recommendation system based on dynamic analysis of customer behavior. *Sustainability, 13*(19), 10786. <https://doi.org/10.3390/su131910786>

Acemoglu, D., & Restrepo, P. (2018). Artificial intelligence, automation and work. *National Bureau of Economic Research.*<https://doi.org/10.3386/w24196>

Aleedy, M., Atwell, E., & Meshoul, S. (2022). Using AI Chatbots in education: Recent advances, challenges and use case. In M. Pandit, M. K. Gaur, P. S. Rana, & A. Tiwari (Eds.), *Artificial intelligence and sustainable computing: Algorithms for intelligent systems* (pp. 663–675). Springer. <https://doi.org/10.1007/978-981-19-1653-3_50>

Algül, A., Yengin, D., Karadağ, G. H., Övür, A., & Bayrak, T. (2018). *Sanal gerçekliğin tetiklediği semptomlar*. İstanbul Aydın Üniversitesi Yayınları.

Amazon. (n.d.). Alexa. Retrieved August 16, 2024, from <https://alexa.amazon.com>

Anantrasirichai, N., & Bull, D. (2022). Artificial intelligence in the creative industries: A review. *Artificial Intelligence Review, 55*, 589–656. <https://doi.org/10.1007/s10462-021-10039-7>

Arnold, K. E., & Pistilli, M. D. (2012). Course signals at Purdue: Using learning analytics to increase student success. In *Proceedings of the 2nd International Conference on Learning Analytics and Knowledge (LAK '12)* (pp. 267–270). Association for Computing Machinery. <https://doi.org/10.1145/2330601.2330666>

BIBLIOGRAPHY

Arute, F., Arya, K., Babbush, R., Bacon, D., Bardin, J. C., Barends, R., Biswas, R., Boixo, S., Brandao, F. G. S. L., Buell, D. A., Burkett, B., Chen, Y., Chen, Z., Chiaro, B., Collins, R., Courtney, W., Dunsworth, A., Farhi, E., Foxen, B., ... Martinis, J. M. (2019). Quantum supremacy using a programmable superconducting processor. *Nature*, 574(7779), 505–510. <https://doi.org/10.1038/s41586-019-1666-5>

Autor, D. H. (2015). Why are there still so many jobs? The history and future of workplace automation. *Journal of Economic Perspectives*, 29(3), 3–30. <https://doi.org/10.1257/jep.29.3.3>

Babatunde, S. O., Odejide, O. A., Edunjobi, T. E., & Ogundipe, D. O. (2024). The role of AI in marketing personalization: A theoretical exploration of consumer engagement strategies. *International Journal of Management & Entrepreneurship Research*, 6(3), 936–949. <https://doi.org/10.51594/ijmer.v6i3.964>

Baethge, C., Klier, J., & Klier, M. (2016). Social commerce—State-of-the-art and future research directions. *Electronic Markets*, 26(3), 269–290. <https://doi.org/10.1007/s12525-016-0225-2>

Bank of America. (n.d.). Erica: Your virtual financial assistant. Retrieved August 16, 2024, from <https://promotions.bankofamerica.com/digitalbanking/mobilebanking/erica>

Bansal, A., Ranjan, R., Castillo, C. D., & Chellappa, R. (2021). Deep CNN face recognition: Looking at the past and the future. In N. K. Ratha, V. M. Patel, & R. Chellappa (Eds.), *Deep learning-based face analytics* (pp. 6–29). Springer. <https://doi.org/10.1007/978-3-030-74697-1_2>

Barocas, S., Hardt, M., & Narayanan, A. (2019). *Fairness and machine learning*. MIT Press.

Bender, E. M., Gebru, T., McMillan-Major, A., & Shmitchell, S. (2021). On the dangers of stochastic parrots: Can language models be too big? In *Proceedings of the 2021 ACM Conference on Fairness, Accountability, and Transparency (FAccT '21)* (pp. 610–623). ACM. <https://doi.org/10.1145/3442188.3445922>

Bernardo, J. M., & Smith, A. F. M. (2000). *Bayesian theory*. John Wiley & Sons.

Binns, R. (2018). Fairness in machine learning: Lessons from political philosophy. In *Proceedings of the 1st Conference on Fairness, Accountability and Transparency*. Proceedings of Machine Learning Research (Vol. 81, pp. 149–159). <https://proceedings.mlr.press/v81/binns18a.html>

Boahen, K. (2017). A neuromorph's prospectus. *Computing in Science & Engineering, 19*(2), 14–28. <https://doi.org/10.1109/MCSE.2017.33>

Bontridder, N., & Poullet, Y. (2021). The role of artificial intelligence in disinformation. *Data & Policy, 3*, e32. <https://doi.org/10.1017/dap.2021.20>

Booking.com. (n.d.). Booking.com. Retrieved August 16, 2024, from <https://www.booking.com>

Bostrom, N. (2014). *Superintelligence: Paths, dangers, strategies*. Oxford University Press.

Bostrom, N., & Yudkowsky, E. (2014). The ethics of artificial intelligence. In K. Frankish & W. Ramsey (Eds.), *The Cambridge handbook of artificial intelligence* (pp. 385–417). Cambridge University Press.

Bourne, C. (2023). AI hype: Public relations and AI's doomsday machine. In A. Adi (Ed.), *Artificial intelligence in public relations and communications: Cases, reflections, and predictions* (pp. 39–50). Quadriga University of Applied Sciences.

Boyd, D., & Crawford, K. (2012). Critical questions for big data. *Information, Communication & Society, 15*(5), 662–679. <https://doi.org/10.1080/1369118X.2012.678878>

Broadbent, E., Stafford, R., & MacDonald, B. (2009). Acceptance of healthcare robots for the older population: Review and future directions. *International Journal of Social Robotics, 1*(4), 319–330. <https://doi.org/10.1007/s12369-009-0030-6>

Brynjolfsson, E., & McAfee, A. (2014). *The second machine age: Work, progress, and prosperity in a time of brilliant technologies*. W.W. Norton & Company.

Built In. (n.d.). What is artificial general intelligence (AGI)? Retrieved August 6, 2024, from <https://builtin.com/artificial-intelligence/artificial-general-intelligence>

Buoy Health. (n.d.). Buoy Health. Retrieved August 16, 2024, from <https://www.buoyhealth.com>

Carpio, R., Baumann, O., & Birt, J. (2023). Evaluating the viewer experience of interactive virtual reality movies. *Virtual Reality, 27*, 3181–3190. <https://doi.org/10.1007/s10055-023-00864-2>

Castells, M. (2010). *The rise of the network society* (2nd ed.). Wiley-Blackwell.

Cath, C., Wachter, S., Mittelstadt, B., Taddeo, M., & Floridi, L. (2017). Artificial intelligence and the 'Good Society': The US, EU, and UK approach. *Science* and *Engineering Ethics*, 24(2), 505–528. <https://doi.org/10.1007/s11948-017-9901-7>

Chaddad, A., Peng, J., Xu, J., & Bouridane, A. (2023). Survey of explainable AI techniques in healthcare. *Sensors*, 23(2), 634. <https://doi.org/10.3390/s23020634>

Chaffey, D., & Ellis-Chadwick, F. (2016). *Digital marketing: Strategy, implementation, and practice* (6th ed.). Pearson Education Limited.

Chancellor, S., Birnbaum, M. L., Caine, E. D., Silenzio, V. M. B., & De Choudhury, M. (2019). A taxonomy of ethical tensions in inferring mental health states from social media. In *Proceedings of the Conference on Fairness, Accountability, and Transparency (FAT '19)* (pp. 79–88). Association for Computing Machinery. <https://doi.org/10.1145/3287560.3287587>

Chandra, S., Verma, S., Lim, W. M., Kumar, S., & Donthu, N. (2022). Personalization in personalized marketing: Trends and ways forward. *Psychology & Marketing*, 39(8), 1529–1562. <https://doi.org/10.1002/mar.21670>

Chapman University. (n.d.). PantherBot. Retrieved August 16, 2024, from <https://www.chapman.edu/campus-services/information-systems/services/pantherbot.aspx>

Chaudhry, M. A., & Kazim, E. (2022). Artificial intelligence in education (AIEd): A high-level academic and industry note 2021. *AI Ethics*, 2(2), 157–165. <https://doi.org/10.1007/s43681-021-00074-z>

Chen, C., Seff, A., Kornhauser, A., & Xiao, J. (2015). DeepDriving: Learning affordance for direct perception in autonomous driving. In *2015 IEEE International Conference on Computer Vision (ICCV)* (pp. 2722–2730). <https://doi.org/10.1109/ICCV.2015.312>

Colby, K. M., Goldstein, A. P., & Krasner, L. (1975). *Artificial paranoia: A computer simulation of paranoid processes*. Pergamon Press.

Connock, A. (2023). *Media management and artificial intelligence: Understanding media business models in the digital age*. Routledge.

Conversation Design Institute. (n.d.). Chatbots for banking. Retrieved August 16, 2024, from <https://www.conversationdesigninstitute.com/blog/chatbots-for-banking>

Covington, P., Adams, J., & Sargin, E. (2016). Deep neural networks for YouTube recommendations. In *Proceedings of the 10th ACM Conference on*

Recommender Systems (RecSys '16) (pp. 191–198). Association for Computing Machinery. <https://doi.org/10.1145/2959100.2959190>

Crawford, K. (2021). *The atlas of AI: Power, politics, and the planetary costs of artificial intelligence*. Yale University Press.

Damassino, N., & Novelli, N. (2020). Rethinking, reworking, and revolutionising the Turing test. *Minds & Machines, 30*(4), 463–468. <https://doi.org/10.1007/s11023-020-09553-4>

Deleuze, G. (1992). Postscript on the societies of control. *October, 59*, 3–7. <http://www.jstor.org/stable/778828>

DevX. (n.d.). Artificial linguistic computer entity. *DevX Technology Glossary*. Retrieved October 12, 2024, from <https://www.devx.com/terms/artificial-linguistic-computer-entity/>

Devlin, J., Chang, M.-W., Lee, K., & Toutanova, K. (2019). BERT: Pre-training of deep bidirectional transformers for language understanding. In *Proceedings of the 2019 Conference of the North American Chapter of the Association for Computational Linguistics: Human Language Technologies, Volume 1 (Long and Short Papers)* (pp. 4171–4186). Association for Computational Linguistics. <https://doi.org/10.18653/v1/N19-1423>

Diakopoulos, N. (2016). Accountability in algorithmic decision-making. *Communications of the ACM, 59*(2), 56–62. <https://doi.org/10.1145/2844110>

Dixit, S., Kumar, A., Srinivasan, K., Vincent, P. M. D. R., & Ramu Krishnan, N. (2024). Advancing genome editing with artificial intelligence: Opportunities, challenges, and future directions. *Frontiers in Bioengineering and Biotechnology, 11*, Article 1335901. <https://doi.org/10.3389/fbioe.2023.1335901>

Domingos, P., & Pazzani, M. (1997). On the optimality of the simple Bayesian classifier under zero-one loss. *Machine Learning, 29*(2–3), 103–130. <https://doi.org/10.1023/A:1007413511361>

Duolingo. (n.d.). Duolingo. Retrieved August 16, 2024, from <https://en.duolingo.com>

El-emary, I. M. M., Fezari, M., & Attoui, H. (2011). Hidden Markov model/Gaussian mixture models (HMM/GMM) based voice command system: A way to improve the control of remotely operated robot arm. *Scientific Research and Essays, 6*(2), 341–350.

Elgammal, A., Liu, B., Elhoseiny, M., & Mazzone, M. (2017). CAN: Creative adversarial networks, generating "art" by learning about styles and deviating from style norms.

Esteva, A., Kuprel, B., Novoa, R., Ko, J., Swetter, S. M., Blau, H. M., & Thrun, S. (2017). Dermatologist-level classification of skin cancer with deep neural networks. *Nature, 542*(7639), 115–118. <https://doi.org/10.1038/nature21056>

European Commission. (2021). Proposal for a regulation of the European Parliament and of the Council laying down harmonized rules on artificial intelligence (Artificial Intelligence Act) and amending certain Union legislative acts.

Feijoo-Garcia, M. A., Holstrom, M. S., Magana, A. J., & Newell, B. A. (2024). Simulation-based learning and argumentation to promote informed design decision-making processes within a first-year engineering technology course. *Sustainability, 16*(7), 2633. <https://doi.org/10.3390/su16072633>

Filipović, A. (2023). The role of artificial intelligence in video game development. *Kultura Polisa, 20*(3), 50–67. <https://doi.org/10.51738/Kpolisa2023.20.3r.50f>

Floridi, L. (2019). Translating principles into practices of digital ethics: Five risks of being unethical. *Philosophy & Technology, 32*(2), 185–193. <https://doi.org/10.1007/s13347-019-00354-x>

Floridi, L. (2023). *The ethics of artificial intelligence: Principles, challenges, and opportunities*. Oxford University Press. <https://doi.org/10.1093/oso/9780198883098.001.0001>

Floridi, L., & Cowls, J. (2019). A unified framework of five principles for AI in society. *Harvard Data Science Review, 1*, 2–15. <https://doi.org/10.1162/99608f92.8cd550d1>

Fong, T., Nourbakhsh, I., & Dautenhahn, K. (2003). A survey of socially interactive robots. *Robotics and Autonomous Systems, 42*(3–4), 143–166. <https://doi.org/10.1016/S0921-8890(02)00372-X>

Frey, C. B., & Osborne, M. A. (2017). The future of employment: How susceptible are jobs to computerisation? *Technological Forecasting and Social Change, 114*, 254–280. <https://doi.org/10.1016/j.techfore.2016.08.019>

Fuchs, C. (2020). *Social media: A critical introduction*. SAGE Publications.

Gambardella, V., Tarazona, N., Cejalvo, J. M., Lombardi, P., Huerta, M., Roselló, S., Fleitas, T., Roda, D., & Cervantes, A. (2020). Personalized medicine: Recent progress in cancer therapy. *Cancers, 12*(4), 1009. <https://doi.org/10.3390/cancers12041009>

Gelman, A., Carlin, J. B., Stern, H. S., Dunson, D. B., Vehtari, A., & Rubin, D. B. (2013). *Bayesian data analysis* (3rd ed.). CRC Press.

Gentsch, P. (2019). *AI in marketing, sales, and service: How marketers without a data science degree can use AI, big data, and bots.* Springer Nature Switzerland AG. <https://doi.org/10.1007/978-3-319-89957-2>

Gillespie, T. (2014). The relevance of algorithms. In T. Gillespie, P. J. Boczkowski, & K. A. Foot (Eds.), *Media technologies: Essays on communication, materiality, and society* (pp. 167–194). MIT Press. <https://doi.org/10.7551/mitpress/9780262525374.003.0009>

Gillespie, T. (2018). *Custodians of the Internet: Platforms, content moderation, and the hidden decisions that shape social media.* Yale University Press.

Gillespie, T. (2020). Content moderation, AI, and the question of scale. *Big Data & Society, 7*(2). <https://doi.org/10.1177/2053951720943234>

Girshick, R., Donahue, J., Darrell, T., & Malik, J. (2014). Rich feature hierarchies for accurate object detection and semantic segmentation. In *2014 IEEE Conference on Computer Vision and Pattern Recognition (CVPR)* (pp. 580–587). <https://doi.org/10.1109/CVPR.2014.81>

Gkikas, D., & Theodoridis, P. (2022). AI in consumer behavior. In M. Virvou, G. A. Tsihrintzis, L. H. Tsoukalas, & L. C. Jain (Eds.), *Advances in artificial intelligence-based technologies.* Learning and Analytics in Intelligent Systems (Vol. 22, pp. 129–141). Springer. <https://doi.org/10.1007/978-3-030-80571-5_10>

Gogpac. (n.d.). How to use AI in your job search. *LinkedIn.* Retrieved August 27, 2024, from https://www.linkedin.com/pulse/how-use-ai-your-job-search-gogpac-po4yc

Gonzales, J. T. (2023). Implications of AI innovation on economic growth: A panel data study. *Economic Structures, 12*(13). <https://doi.org/10.1186/s40008-023-00307-w>

Gonzalez, R. C., & Woods, R. E. (2002). *Digital image processing* (2nd ed.). Prentice Hall.

Goodfellow, I., Bengio, Y., & Courville, A. (2016). *Deep learning.* MIT Press.

Gordon, F. (2019). Virginia Eubanks (2018) Automating inequality: How high-tech tools profile, police, and punish the poor. *Law, Technology and Humans, 1*, 162–164. <https://doi.org/10.5204/lthj.v1i0.1386>

Grandinetti, J. (2021). Examining embedded apparatuses of AI in Facebook and TikTok. *AI & Society, 38*(3), 1273–1286. <https://doi.org/10.1007/s00146-021-01270-5>

Graves, A., Mohamed, A.-r., & Hinton, G. (2013). Speech recognition with deep recurrent neural networks. In *2013 IEEE International Conference on Acoustics, Speech and Signal Processing* (pp. 6645–6649). IEEE. <https://doi.org/10.1109/ICASSP.2013.6638947>

Habes, M., Alhazmi, A. H., Elareshi, M., & Attar, R. W. (2024). Understanding the relationship between AI and gender on social TV content selection. *Frontiers in Communication, 9*, Article 1410995. <https://doi.org/10.3389/fcomm.2024.1410995>

Haenlein, M., & Kaplan, A. (2019). A brief history of artificial intelligence: On the past, present, and future of artificial intelligence. *California Management Review, 61*(4), 5–14. <https://doi.org/10.1177/0008125619864925>

Hakimi, M., Sazish, B., Rastagari, M. A., & Shahidzay, A. K. (2023). Artificial intelligence for social media safety and security: A systematic literature review. *Studies in Media, Journalism, and Communications, 1*(1), 10–21. <https://doi.org/10.32996/smjc.2023.1.1.2x>

Hameleers, M., & van der Meer, T. G. L. A. (2020). Misinformation and polarization in a high-choice media environment: How effective are political fact-checkers?. *Communication Research, 47*(2), 227–250. <https://doi.org/10.1177/0093650218819671>

Harry, A. (2023). The future of medicine: Harnessing the power of AI for revolutionizing healthcare. *International Journal of Multidisciplinary Sciences and Arts, 2*(1), 36–44. <https://doi.org/10.47709/ijmdsa.v2i1.2395>

Harvard Science Review. (n.d.). Artificial superintelligence: The coming revolution. Retrieved August 6, 2024, from <https://harvardsciencereview.com/artificial-superintelligence-the-coming-revolution/>

Hinton, G., Deng, L., Yu, D., Dahl, G. E., Mohamed, A. R., Jaitly, N., Senior, A., Vanhoucke, V., Nguyen, P., Sainath, T. N., & Kingsbury, B. (2012). Deep neural networks for acoustic modeling in speech recognition: The shared views of four research groups. *IEEE Signal Processing Magazine, 29*(6), 82–97. <https://doi.org/10.1109/MSP.2012.2205597>

Holmes, W., Bialik, M., & Fadel, C. (2019). *Artificial intelligence in education: Promises and implications for teaching and learning.* Center for Curriculum Redesign.

Huang, M. H., & Rust, R. T. (2018). Artificial intelligence in service. *Journal of Service Research, 21*(2), 155–172. <https://doi.org/10.1177/1094670517752459>

HubSpot. (n.d.). AI in social media marketing. *HubSpot*. Retrieved August 27, 2024, from <https://www.hubspot.com/products/artificial-intelligence/ai-social>

Interaction Design Foundation - IxDF. (2023, November 20). What is Narrow AI? *Interaction Design Foundation - IxDF*. <https://www.interaction-design.org/literature/topics/narrow-ai>

Jelinek, F. (1997). *Statistical methods for speech recognition*. MIT Press.

Jensen, F. V. (1996). *An introduction to Bayesian networks*. UCL Press.

Jha, A. K., & Verma, N. K. (2024). Social media platforms and user engagement: A multi-platform study on one-way firm sustainability communication. *Information Systems Frontiers, 26*(1), 177–194. <https://doi.org/10.1007/s10796-023-10376-8>

Jurafsky, D., & Martin, J. H. (2023). *Speech and language processing: An introduction to natural language processing, computational linguistics, and speech recognition* (3rd ed., draft). Stanford University and University of Colorado at Boulder.

Kamilaris, A., & Prenafeta-Boldú, F. X. (2018). Deep learning in agriculture: A survey. *Computers and Electronics in Agriculture, 147*, 70–90. <https://doi.org/10.1016/j.compag.2018.02.016>

Kietzmann, J., Paschen, J., & Treen, E. (2018). Artificial intelligence in advertising: How marketers can leverage artificial intelligence along the consumer journey. *Journal of Advertising Research, 58*(3), 263–267. <https://doi.org/10.2501/JAR-2018-035>

Kitsios, F., Kamariotou, M., Syngelakis, A. I., & Talias, M. A. (2023). Recent advances of artificial intelligence in healthcare: A systematic literature review. *Applied Sciences, 13*(7479). <https://doi.org/10.3390/app13137479>

KLM Royal Dutch Airlines. (2017, September 27). KLM launches Messenger customer chat on KLM.com. Retrieved August 16, 2024, from <https://news.klm.com/klm-launches-messenger-customer-chat-on-klmcom/>

Koller, D., & Friedman, N. (2009). *Probabilistic graphical models: Principles and techniques*. MIT Press.

Korinek, A., & Stiglitz, J. E. (2017). Artificial intelligence and its implications for income distribution and unemployment. *NBER Working Paper No. 24174*. <https://doi.org/10.3386/w24174>

Krizhevsky, A., Sutskever, I., & Hinton, G. E. (2012). Imagenet classification with deep convolutional neural networks. *Advances in Neural Information Processing Systems*, 25, 1097–1105.

Kumar, Y. (2024). A comprehensive analysis of speech recognition systems in healthcare: Current research challenges and future prospects. *SN Computer Science*, 5(137). <https://doi.org/10.1007/s42979-023-02466-w>

Kurzweil, R. (2005). *The singularity is near: When humans transcend biology*. Viking.

LeCun, Y., Bengio, Y., & Hinton, G. (2015). Deep learning. *Nature*, 521(7553), 436–444. <https://doi.org/10.1038/nature14539>

Levi Strauss & Co. (2017, August 31). Levi's launches new virtual stylist online feature. Retrieved August 16, 2024, from <https://www.levistrauss.com/2017/08/31/levis-launches-new-virtual-stylist-online-feature/>

Li, F., Ruijs, N., & Lu, Y. (2023). Ethics & AI: A systematic review on ethical concerns and related strategies for designing with AI in healthcare. *AI*, 4(1), 28–53. <https://doi.org/10.3390/ai4010003>

Liang, C., Chou, W.-S., Hsu, Y.-L., & Yang, C.-C. (2009). A user-centered design approach to develop a web-based instructional resource system for homeland education. *Knowledge Management & E-Learning: An International Journal*, 1(1), 67–80.

Liang, Q. (2024). Automatic speech recognition technology: History, applications and improvements. *Applied and Computational Engineering*, 65, 180–184.

Litjens, G., Kooi, T., Ehteshami Bejnordi, B., Setio, A. A. A., Ciompi, F., Ghafoorian, M., van der Laak, J. A. W. M., van Ginneken, B., & Sánchez, C. I. (2017). A survey on deep learning in medical image analysis. *Medical Image Analysis*, 42, 60–88. <https://doi.org/10.1016/j.media.2017.07.005>

Lopatin, J., Dolos, K., Hernández, H. J., Galleguillos, M., & Fassnacht, F. E. (2016). Comparing generalized linear models and random forest to model vascular plant species richness using LiDAR data in a natural forest in central Chile. *Remote Sensing of Environment*, 173, 200–210. <https://doi.org/10.1016/j.rse.2015.11.029>

López Jiménez, E. A., & Ouariachi, T. (2021). An exploration of the impact of artificial intelligence (AI) and automation for communication professionals. *Journal of Information, Communication and Ethics in Society*, 19(2), 249–267. <https://doi.org/10.1108/JICES-03-2020-0034>

Lund, B., & Ghiasi, N. (2024). Ethical considerations in artificial intelligence interventions for mental health and well-being: Ensuring responsible implementation and impact. *Social Sciences, 13*(7), 381. <https://doi.org/10.3390/socsci13070381>

Manning, C. D., & Schütze, H. (1999). *Foundations of statistical natural language processing.* MIT Press.

McCorduck, P. (2004). *Machines who think* (2nd ed.). A. K. Peters, Ltd.

Mention. (2024). 7 ways LinkedIn's AI is transforming professional networking. *Mention.* Retrieved August 27, 2024, from <https://mention.com/en/blog/linkedin-ai/>

Mitchell, T. M. (1997). *Machine learning.* McGraw-Hill.

Modernizing the Turing Test for 21st Century AI. (2021). *The Next Platform.* Retrieved from NextPlatform.

Mohammadzadeh, A., Sabzalian, M. H., Castillo, O., Sakthivel, R., El-Sousy, F. F. M., & Mobayen, S. (2022). Multilayer perceptron (MLP) neural networks. In *Neural networks and learning algorithms in MATLAB* (pp. 45-62). Synthesis Lectures on Intelligent Technologies. Springer, Cham. <https://doi.org/10.1007/978-3-031-14571-1_2>

Murphy, K. P. (2012). *Machine learning: A probabilistic perspective.* MIT Press.

Nagaraj, B. K. (2022). Integration of AI and neuroscience for advancing brain-machine interfaces: A study. *International Journal of New Media Studies, 9*(1), 25–30.

Nancholas, B. (2023, September 1). Narrow artificial intelligence: Advantages, disadvantages, and the future of AI. *University of Wolverhampton.* <https://online.wlv.ac.uk/narrow-artificial-intelligence-advantages-disadvantages-and-the-future-of-ai/>

Neuendorf, K. A. (2017). *The content analysis guidebook* (2nd ed.). SAGE Publications.

NYC Data Science Academy. (n.d.). Predicting house prices using machine learning: What features matter most? *NYC Data Science Academy.* Retrieved August 3, 2024, from <https://nycdatascience.com/blog/student-works/predicting-house-prices-using-machine-learning-what-features-matter-most/>

O'Neil, C. (2016). *Weapons of math destruction: How big data increases inequality and threatens democracy.* Crown Publishers.

Paredes, W. C., & Chung, K. S. K. (2012). Modelling learning & performance: A social networks perspective. In *Proceedings of the 2nd International Conference on Learning Analytics and Knowledge (LAK '12)* (pp. 34–42). Association for Computing Machinery. <https://doi.org/10.1145/2330601.2330617>

Pardo, A., & Siemens, G. (2014). Ethical and privacy principles for learning analytics. *British Journal of Educational Technology, 45*(3), 438–450. <https://doi.org/10.1111/bjet.12152>

Pariser, E. (2011). *The filter bubble: What the internet is hiding from you*. Penguin Press.

Pasquale, F. (2015). *The black box society: The secret algorithms that control money and information*. Harvard University Press.

Patra, B. G., Sharma, M. M., Vekaria, V., Adekkanattu, P., Patterson, O. V., Glicksberg, B., Lepow, L. A., Ryu, E., Biernacka, J. M., Furmanchuk, A., George, T. J., Hogan, W., Wu, Y., Yang, X., Bian, J., Weissman, M., Wickramaratne, P., Mann, J. J., Olfson, M., Campion, T. R., Weiner, M., & Pathak, J. (2021). Extracting social determinants of health from electronic health records using natural language processing: A systematic review. *Journal of the American Medical Informatics Association, 28*(12), 2716–2727. <https://doi.org/10.1093/jamia/ocab170>

Pearl, J. (1988). *Probabilistic reasoning in intelligent systems: Networks of plausible inference*. Morgan Kaufmann.

Picard, R. W. (2010). Affective computing: From laughter to IEEE. *IEEE Transactions on Affective Computing, 1*(1), 11–17. <https://doi.org/10.1109/t-affc.2010.10>

Radziwill, N. M., & Benton, M. C. (2017). Evaluating quality of chatbots and intelligent conversational agents. *Journal of Computer Information Systems, 58*(1), 1–12. <https://doi.org/10.1080/08874417.2017.1373077>

Reddy, S., Fox, J., & Purohit, M. P. (2019). Artificial intelligence-enabled healthcare delivery. *Journal of the Royal Society of Medicine, 112*(1), 22–28. <https://doi.org/10.1177/0141076818815510>

Redmon, J., Divvala, S., Girshick, R., & Farhadi, A. (2016). You only look once: Unified, real-time object detection. In *Proceedings of the IEEE Conference on Computer Vision and Pattern Recognition* (pp. 779–788). <https://doi.org/10.1109/CVPR.2016.91>

Refik Anadol Studio. (n.d.). Refik Anadol. Retrieved September 2, 2024, from <https://refikanadol.com>

Republic of Türkiye Presidency Digital Transformation Office. (2021). National artificial intelligence strategy 2021–2025. Retrieved August 28, 2024, from <https://bit.ly/3Mniubj>

Rodrigues, A. P., Fernandes, R., Bhandary, A., Shenoy, A. C., Shetty, A., & Anisha, M. (2021). Real-time Twitter trend analysis using big data analytics and machine learning techniques. *Wireless Communications and Mobile Computing*, Article ID 3920325. <https://doi.org/10.1155/2021/3920325>

Roketto. (n.d.). Automated content generation: What it means for marketers. *Roketto*. Retrieved August 27, 2024, from <https://www.helloroketto.com/articles/automated-content-generation>

Rosário, A. T., & Dias, J. C. (2023). Marketing strategies on social media platforms. *International Journal of E-Business Research*, *19*(1). <https://doi.org/10.4018/IJEBR.316969>

Russell, S., & Norvig, P. (2010). *Artificial intelligence: A modern approach* (3rd ed.). Prentice Hall.

Quazi, S., Saha, R., & Kumar, M. (2022). Applications of artificial intelligence in healthcare. *Journal of Experimental Biology and Agricultural Sciences*, *10*(1), 211–226. <https://doi.org/10.18006/2022.10(1).211.226>

Sadiku, M. N. O., Ashaolu, T. J., Ajayi-Majebi, A., & Musa, S. M. (2021). Artificial intelligence in social media. *International Journal of Scientific Advances*, *2*(1), 15–20. <https://doi.org/10.51542/ijscia.v2i1.4>

Salas-Pilco, S. Z., Xiao, K., & Hu, X. (2022). Artificial intelligence and learning analytics in teacher education: A systematic review. *Education Sciences*, *12*(8), 569. <https://doi.org/10.3390/educsci12080569>

Sarker, I. H. (2021). Machine learning: Algorithms, real-world applications and research directions. *SN Computer Science*, *2*, 160. <https://doi.org/10.1007/s42979-021-00592-x>

Schneier, B. (2015). *Data and Goliath: The hidden battles to collect your data and control your world*. W.W. Norton & Company.

Seidenglanz, R., & Baier, M. (2023). The impact of artificial intelligence on the professional field of public relations/communications management: Recent developments and opportunities. In A. Adi (Ed.), *Artificial intelligence in public relations and communications: Cases, reflections, and predictions* (pp. 14–25). Quadriga University of Applied Sciences.

Sephora. (n.d.). Virtual artist. Retrieved August 16, 2024, from <https://www.sephora.sg/pages/virtual-artist>

Szeliski, R. (2022). *Computer vision: Algorithms and applications* (2nd ed.). Springer.

Shen, Y.-T., Chen, L., Yue, W.-W., & Xu, H.-X. (2021). Artificial intelligence in ultrasound. *European Journal of Radiology, 139*, 109717. <https://doi.org/10.1016/j.ejrad.2021.109717>

Shum, H., He, X., & Li, D. (2018). From ELIZA to XiaoIce: Challenges and opportunities with social chatbots. *Frontiers of Information Technology & Electronic Engineering, 19*(1), 10–26. <https://doi.org/10.1631/FITEE.1700826>

Tegmark, M. (2017). *Life 3.0: Being human in the age of artificial intelligence*. Alfred A. Knopf.

Tepe, R. (2020). What is the role of HCI within the automotive industry? (Bachelor's thesis, Newcastle College University Centre). Retrieved from <https://tls.tc/Erkdn>

The History of Artificial Intelligence. (2023). *Science in the News*. Harvard University. Retrieved from Science in the News.

Tolba, A. S., El-Baz, A. H., & El-Harby, A. A. (2005). Face recognition: A literature review. *International Journal of Signal Processing, 2*(2), 88–98.

Topol, E. (2019). *Deep medicine: How artificial intelligence can make healthcare human again*. Basic Books.

Trotta, A., Ziosi, M., & Lomonaco, V. (2023). The future of ethics in AI: Challenges and opportunities. *AI & Society, 38*. <https://doi.org/10.1007/s00146-023-01644-x>

Turing, A. M. (1950). Computing machinery and intelligence. *Mind, 59*(236), 433–460. Retrieved from Mind Journal.

Turkle, S. (2011). *Alone together: Why we expect more from technology and less from each other*. Basic Books.

UNESCO. (2021). *Recommendation on the ethics of artificial intelligence*. UNESCO. <https://www.unesco.org/en/artificial-intelligence/recommendation-ethics>

van der Maaten, L., & Hinton, G. (2008). Visualizing data using t-SNE. *Journal of Machine Learning Research, 9*, 2579–2605.

van Dijck, J. (2013). *The culture of connectivity: A critical history of social media*. Oxford University Press.

van Dijck, J., Poell, T., & de Waal, M. (2018). *The platform society: Public values in a connective world*. Oxford University Press.

Van Krevelen, D., & Poelman, R. (2010). A survey of augmented reality technologies, applications and limitations. *International Journal of Virtual Reality*, 9(2), 1–20. <https://doi.org/10.20870/IJVR.2010.9.2.2767>

Vaswani, A., Shazeer, N., Parmar, N., Uszkoreit, J., Jones, L., Gomez, A. N., Kaiser, L., & Polosukhin, I. (2017). Attention is all you need. *Advances in Neural Information Processing Systems*, 30, 5998–6008.

Vollero, A., Sardanelli, D., & Siano, A. (2021). Exploring the role of the Amazon effect on customer expectations: An analysis of user-generated content in consumer electronics retailing. *Journal of Consumer Behaviour*, 22(5), 1062–1073. <https://doi.org/10.1002/cb.1969>

Weizenbaum, J. (1966). ELIZA—a computer program for the study of natural language communication between man and machine. *Communications of the ACM*, 9(1), 36–45. <https://doi.org/10.1145/365153.365168>

Wilding, D., Fray, P., Molitorisz, S., & McKewon, E. (2018). The impact of digital platforms on news and journalistic content. University of Technology Sydney.

Woebot Health. (n.d.). Woebot Health. Retrieved August 16, 2024, from <https://woebothealth.com>

Wu, S., Huang, S., Chen, W., Xiao, F., & Zhang, W. (2022). Design and implementation of intelligent car controlled by voice. In *2022 International Conference on Computer Network, Electronic and Automation (ICCNEA)* (pp. 326–330). IEEE. <https://doi.org/10.1109/ICCNEA57056.2022.00078>

Wu, Y., Schuster, M., Chen, Z., Le, Q. V., Norouzi, M., Macherey, W., Krikun, M., Cao, Y., Gao, Q., Macherey, K., & others. (2016). Google's neural machine translation system: Bridging the gap between human and machine translation.

Xiong, W., Droppo, J., Huang, X., Seide, F., Seltzer, M., Stolcke, A., Yu, D., & Zweig, G. (2017). Toward human parity in conversational speech recognition. *IEEE/ACM Transactions on Audio, Speech, and Language Processing*, 25(12), 2410–2423. <https://doi.org/10.1109/TASLP.2017.2756440>

Yengin, D., & Bayrak, T. (2017). *Sanal gerçeklik - VR*. Der Yayınevi.

Yengin, D., & Bayrak, T. (2023). *Yeni medya kuram ve yaklaşımlar 101*. Der Yayınevi.

Yudkowsky, E. (2008). Artificial intelligence as a positive and negative factor in global risk. In N. Bostrom & M. M. Ćirković (Eds.), *Global catastrophic risks* (pp. 308–345). Oxford University Press.

Zawacki-Richter, O., Marín, V. I., Bond, M., & Gouverneur, F. (2019). Systematic review of research on artificial intelligence applications in higher education – Where are the educators? *International Journal of Educational Technology in Higher Education, 16*(1), 39. <https://doi.org/10.1186/s41239-019-0171-0>

Zhou, L., DeAlmeida, D., & Parmanto, B. (2019). Applying a user-centered approach to building a mobile personal health record app: Development and usability study. *JMIR mHealth and uHealth, 7*(7), e13194. <https://doi.org/10.2196/13194>

Zhao, W., Chellappa, R., Phillips, P. J., & Rosenfeld, A. (2003). Face recognition: A literature survey. *ACM Computing Surveys, 35*(4), 399–458. <https://doi.org/10.1145/954339.954342>

Zuboff, S. (2019). *The age of surveillance capitalism: The fight for a human future at the new frontier of power*. PublicAffairs.